OCCUPATIONAL HEALTH

Edited by **Orhan Korhan**

Occupational Health
http://dx.doi.org/10.5772/63281
Edited by Orhan Korhan

Contributors

Nilgun Ulutasdemir, Ferdi Tanır, Marina Otelea, Marcello Silvae Santos, Bernardo Bastos da Fonseca, Ana Cláudia Coelho, Radostina A. Angelova, Simona Šarotar Žižek, Mohamed Mokdad, Tawfik Abdel Moniem, Ari-Matti Auvinen, Ann E Barr-Gillespie, Mary F Barbe, Azra Huršidić Radulović, Rosana Quintana, Vita Dolžan, Alenka Franko, Metoda Dodic-Fikfak

Published by InTech

Janeza Trdine 9, 51000 Rijeka, Croatia

Publishing Process Manager Martina Usljebrka

Technical Editor SPi Global

Cover InTech Design team

First published February, 2017

Printed in Croatia

Additional hard copies can be obtained from orders@intechopen.com

Occupational Health, Edited by Orhan Korhan

p. cm.

Print ISBN 978-953-51-2889-2
Online ISBN 978-953-51-2890-8

For EU product safety concerns:
IN TECH d.o.o., Prolaz Marije Krucifikse Kozulić 3, 51000 Rijeka, Croatia,
info@intechopen.com or visit our website at intechopen.com.

Meet the editor

Orhan Korhan graduated with a BS degree from the Eastern Mediterranean University in 2000, an MS degree from the University of Louisville in 2002, and a PhD from the Eastern Mediterranean University in Industrial Engineering in 2010. He has been working at the Eastern Mediterranean University since 2009. He became an assistant professor in 2010 and associate professor in 2014. He is a board member of the Continuing Education Center and currently positioned as an academic affairs coordinator at the Rector's Office. He has been assigned to scientific committee of several international conferences and published several books, book chapters, and papers in various countries. His current research interests include work-related musculoskeletal disorders, educational ergonomics, human-computer interaction, and facilities planning and design.

Contents

Preface

Occupational Health is one of the book titles proposed by InTech in early 2016. I got very excited and thrilled when I got the invitation to edit this book. It was a dream to collect various manuscripts from scientists throughout the world to discuss occupational health issues in several countries. InTech's invitation made this dream come true.

This book covers topics from health and safety management, occupational medicine, work-related musculoskeletal disorders, and occupational protection. Thus, it can be utilized as a guide to identify and analyze hazards, assess risk, apply risk reduction strategies, and manage process safety for various occupations.

In these chapters, you will find all aspects of health and safety in the workplace. To this extent, a wide array of workplace hazards that include physical factors, adverse ergonomic conditions, chemicals, biological agents, and a complex network of safety risks has been discussed extensively.

The first step to cover was to select and disseminate the keywords, which helped us to receive numerous abstracts. These abstracts were then carefully evaluated, and those that were believed to contribute significantly to the literature were invited to take their place in this book. It took several months to edit each and every word of the sent chapters. It was a very fruitful feedback process that we had with our authors.

In this book, you will read chapters from authors in Argentina, Bahrain, Brazil, Bulgaria, Croatia, Finland, India, Portugal, Romania, Slovenia, Turkey, and the USA. They discussed the occupational health issues from their perspectives and reflected and provided solution to the problems experienced from their own eyes. Thus, it makes the contents of this book interest-ing and very valuable.

I, hereby, would like to thank Ms. Martina Usljebrka who helped me with positive attitude at every single step of the publication process. Moreover, I would like to thank my dear wife, Bakiye Yalınç, for her support and encouragement.

Orhan Korhan, PhD
Eastern Mediterranean University, Famagusta,
Turkish Republic of Northern Cyprus (TRNC), Turkey

Section 1

Health and Safety Management

Chapter 1

New Paradigms in Ergonomics: The Positive Ergonomics

Mohamed Mokdad and Tawfik Abdel-Moniem

Additional information is available at the end of the chapter

http://dx.doi.org/10.5772/67064

Abstract

This chapter aims look at ergonomics from a positive point of view. According to International Ergonomics Association, ergonomics is "the scientific discipline concerned with the understanding of interactions among humans and other elements of a system, and the profession that applies theory, principles, data and methods to design in order to optimize human well-being and overall system performance". The major types of ergonomics. Some of them are physical, cognitive, and positive ergonomics. **Positive ergonomics**: Positive ergonomics refers to a new type of ergonomics that stresses the positive aspects of the man-machine system. Its major interest is to make "human-machine system" enjoyable where the human feels pleasant. **Emotional ergonomics**: Similar to positive ergonomics, emotional ergonomics refers to a type of ergonomics that pays attention to the emotional aspects of the man-machine system. **Spiritual ergonomics**: Spiritual ergonomics is based on the idea that spirit is a key factor which determines the employee's health and success in the man machine system, no matter what he/she is doing in that system. **New approach to Occupational health**: When considering the legacy of occupational health, we find that two approaches were adopted throughout of its history. These are: professional harmonization and ergonomics approaches.

Keywords: positive ergonomics, positive design, emotional ergonomics, emotional design, spiritual ergonomics, spiritual design

1. Introduction

Since that man was found on the ground, he realized the value of work. Work occupies an important place in human life. It is a key field after school where individual capabilities, competencies, and skills are shown. Also, it may be the main source from which human beings get the financial resources they need for daily life. Similarly, the place of work is where man lives for many years (almost half of his life). In addition, it is the best space in which the individual meets with others and builds different social relations. Equally, work puts the individual in a particular socio-economic level. Finally, work may be a source of happiness or misery. If the work is corresponding with man's knowledge, skills, abilities, and inclinations and hopes, it is a source of happiness, pleasure, and joy. On the other hand, if disharmony is seen between work and man's abilities, hopes, inclinations, and ambitions, it becomes a source of misery. Finally, work is the only way for production. You cannot imagine that the trees bear fruit without being maintained by man, and the companies, factories, and industrial institutions do not have production without the work of both employers and employees [1].

To make the work a source of happiness, and to avoid making it a source of misery, work has to be done accompanied by the following conditions:

- Assuring work is ergonomically designed.
- Coupling work with conviction and faith.
- Assuring mastery of work.
- Getting rid of hypocrisy at work.
- Putting the necessary efforts work needs.
- Assuring continuity at work.
- Practicing continuous evaluation to work.

Various authors have previously mentioned that ergonomics, especially positive ergonomics, can make workplace a place where employees feel comfortable, happy, calm, and confident with an increased ability to grow and innovate. In addition, they will experience an improved health, greater well-being, promoted excellence, and a better quality of life [2–4]. When workers are having this positive spirit, productivity will most probably flourish and increase. Therefore, it is advised that positive ergonomics should be studied and implemented in the near future [5].

Motivation for the research: The present research was done to shed light on the importance of positiveness in the evolution of ergonomics. It aims to show that positive ergonomics helps to make individuals happier at work. Researchers have performed a lot of research, but in the areas of physical, cognitive, and organizational ergonomics, but little attention has been given to the positive ergonomics. Martino and Morris [4] mentioned that ergonomic thinking has been in some respects positive thinking, if we take into consideration that ergonomics was and still seeking to design or redesign equipment, devices, and processes. In fact, the design goal is a part of positive ergonomics.

Problem statement: Positive psychology specialists [6–8] have for long time suggested that the application of positive psychology principles makes workers happy at work place. Even though positive psychologists state that it can improve workers' happiness in the workplace, but it is an appeal that attains the improvement from within the individual (attitudes, trends, the good personality features, etc.), as happiness is an internal state that does not have strong relationship with external factors. Thus, the positive ergonomics approach which takes into account both individual and context factors affecting work, can to a large extent bring happiness to the workplace. Therefore, this chapter debates the positive ergonomics.

Research questions:: The chapter aims to answer the following questions:

- What is ergonomics?
- What is positive ergonomics?
- What is the relationship of positive ergonomics with occupational health?

2. Approach

A literature search was conducted, using a variety of keywords (ergonomics, positiveness, positive ergonomics, positive design, emotional ergonomics, emotional design, spiritual ergonomics, and spiritual design). This search was conducted as follows: Initially, an electronic search in various databases such as Ergonomics Abstracts, Scopus, ScienceDirect, and PsycINFO was conducted. In addition, manual searches were carried out in the majority of ergonomics journals: Applied Ergonomics, Ergonomics, International Journal of Industrial Ergonomics, International Journal of Human Computer Studies, Theoretical Issues in Ergonomics Science, New Technology, Work and Employment, Reviews of Human Factors and Ergonomics, Human Factors and Ergonomics in Manufacturing, and Le Travail Humain. The result was a number of scientific papers and books were collected (see references) on which this survey research is based.

3. Results

3.1. What is ergonomics?

Ergonomics is a multidisciplinary science. Therefore, there is no single definition of ergonomics. The reader may find it difficult to choose the right definition. Researchers [9–11] have attempted to confine multiple existing definitions. They also tried to extract the contents of these definitions. Some of the main results were: in the years before 1970, the big focus of ergonomics was on effectiveness. In the years between 1970 and 1979, the focus was still on effectiveness and to some extent on comfort and efficiency. In the years after 1980, the focus is still on effectiveness and other factors, such as safety and usability.

Human Factors and Ergonomics Society [12] introduced four criteria which can be used to choose the most comprehensive definition.

1. Does the definition reflect the breadth of human factors and ergonomics?
2. Does the definition comprise both research (theory) and design (application)?
3. Does the definition reflect the scientific nature of the field?
4. Is the definition clear and concise?

In the light of these criteria, two definitions are presented. First, Chapanis [13] wrote on human factors and ergonomics: "Both are concerned with designing for human use and apply information about human characteristics, capacities, and limitations to the design of human tasks, machines, machine systems, and environments so that people can work safely, comfortably, and effectively". After about 45 years, International Ergonomics Association [1] wrote on ergonomics: "Ergonomics (or human factors) is the scientific discipline concerned with the understanding of interactions among humans and other elements of a system, and the profession that applies theory, principles, data and methods to design in order to optimize human well-being and overall system performance" [1].

Looking at these two definitions from the angle of what this chapter advocates, the two definitions, the old and the new, have not clearly stressed the new development in ergonomics (positive, emotional, and spiritual ergonomics).

Development of ergonomics: Talking about the history of ergonomics, one ought to differentiate between non-official and official histories. As to the non-official level, since the man was found on the surface of the earth, he has been designing the simple tools he used in his life, such as knives, spears, and swords to fit his abilities and capacities so that they are used effectively. However, as to the official level, it is believed that ergonomics was officially started in 1949. In a meeting of the British Admiralty, Prof. Hywell Murrell (1908–1984) proposed the name ergonomics that was officially accepted in 1950. Since then, the concept of ergonomics was used.

The development of official ergonomics can be attributed to a great extent to Second World War.

- During the war, researchers through analyzing frequent aircraft accidents, which were piloted by experienced pilots, discovered that accidents are caused by the mistakes of the pilots as well as the bad design of the displays and controls of the aircraft.
- After the war, the impact on countries, especially the allies, was very heavy with great deaths and destruction. Reconstruction has to be done quickly. Authorities have known that the mission needs a lot of efforts and facilities. Ergonomics that emerged almost after the Second World War was one of the different sciences that have to participate in the construction of affected countries.

During a period of over 60 years, ergonomics has greatly developed. The major developments are:

- At the methodological level: Despite the fact that in the early years of ergonomics, most research was using quantitative methods as they are more objective methods [14]. In the twenty-first century, ergonomists felt that there are some ergonomics topics where qualitative methods should be used. Then, a new methodology was widely constituted [15–17].
- At the subject matter level, new types of ergonomics have come into being. They are cognitive ergonomics [18, 19], positive ergonomics [20, 21], emotional ergonomics [22–25], and spiritual ergonomics [26, 27].

Importance of ergonomics: Ergonomics is about both the individual and production. When efficiently applied, ergonomics benefits both human health and company economy.

Human health: Investing in ergonomics increases the health and wellbeing of personnel. Effective ergonomic practices reduce work injuries, for example, carpal tunnel syndrome, tendonitis, and low back pain. These injuries are now considered to be among the leading causes for disability in the modern workplace [28, 29]. In USA, Occupational Safety & Health Administration (OSHA) estimates that as many as 1.8 million work-related musculoskeletal disorders occur every year. These injuries result in a loss of 650,000 work days per year; more than one-third of the total amount of workdays that are lost on an annual basis [30]. Anton and Weeks [31] believe that approximately 80% of participants reported work-related musculoskeletal symptoms. Furthermore, Friedman et al. [32] mention that workplace amputation is a widespread, disabling, costly, and preventable public health problem, with thousands of occupational amputations occurring each year among American workers. Ergonomic practices reduce fatigue. If an employee is less fatigued, he/ she is also less likely to be injured on the job, reducing both absenteeism and the risk of insurance claims. Further, they reduce indirect costs resulting from failure to apply ergonomics in the workplace. When an employee gets injured at work it generates numerous indirect costs such as the payment of the accident expenses, the cost of having some workers log overtime to cover for a missing worker, the decreased morale of employees when knowing a colleague is injured, and the legal and investigation costs if the case reaches the court. Also, they reduce instances of absenteeism due to work-related injury or illness. Similarly, they help them to do more work because they need to take less time to rest.

Besides, ergonomic practices increase the morale of the workers. When workers feel they are being interested in, their morale is boosted, they like the work environment, and most probably, their relationships become stronger.

Company economics: Effective ergonomic practices can increase productivity by reducing the time taken to complete daily tasks. As a result, company profits can also be increased. In addition to the productivity, the quality of production increases. On the other hand, if ergonomics is excluded, most probably, workplaces will cost business a lot of money. In a significant study, Tompa et al. [33] concluded that ergonomic interventions were worth undertaking on the basis of their financial merits. They assisted to reduce frequency or severity of injuries, which eventually result in productivity improvements that result in savings.

Types of ergonomics: Similar to all science, ergonomics, when it originated, was a limited field, but after some years of research, its field expanded to include various subjects. These days, researchers are talking about multiple types of ergonomics. But it would be a good idea to classify these types. The basis of classification would be:

- Classification based upon body part: physical, cognitive, emotional, and spiritual ergonomics.
- Classification based upon application area: educational, industrial, agricultural, and service ergonomics.
- Classification based upon interest area: traditional and transfer of technology ergonomics.
- Classification based upon level of intervention: organizational and individual ergonomics.

- Classification based upon type of intervention: positive and negative ergonomics.
- Classification based upon subject of intervention: worker-centered and production-centered ergonomics.

It should be noted that the most widely used classification in many ergonomics books and literature is the first one (physical, cognitive, and organizational). However, the problem with this classification is not exhaustive as it excludes emotional and spiritual types of ergonomics and includes another type (organizational ergonomics) that belongs to another classification. It is well-known that a good classification is one that is stable, flexible, exhaustive, and having mutually exclusive classes [34].

3.2. What is positive ergonomics?

Around the mid of the last century, psychology researchers put forward the idea that psychology in addition to studying the negative aspects of human personality should also, study the positive ones. In these instances, the positive psychology movement emerged. According to Gable and Haidt [35], positive psychology is the study of the conditions and processes that contribute to the flourishing or optimal functioning of people, groups, and institutions. Previously researchers were more interested in the negative aspects of human personality such as depression, violence, racism, and pessimism, but, less interested in positive aspects such as joy, peace, altruism, humbleness, and optimism. The major aim was not only to return human being to normal, but also to develop different abilities and strengths. The effects of this approach (paradigm shift) did not remain confined to the field of psychology, but surpassed them to the field of ergonomics. In the last few years, researchers began to talk about the positive ergonomics, and the concept of hedonomics appeared.

Definition: Positive ergonomics refers to a new type of ergonomics that stresses the positive aspects of the man-machine system. Previously, ergonomics used to make sure that man-machine system, especially the complex one, is functioning well. Therefore, before positive ergonomics, what was intended from man-machine systems is functionality; however, in the era of positive ergonomics, attention was shifted to pleasurablility. The major mission of positive ergonomics is to understand and foster the factors that allow man-machine systems to flourish. The major interest of positive ergonomics is to make "human-machine system" enjoyable where the human feels pleasant. Traditional ergonomics aimed to make man-machine system effective so that man will not make work accidents, will not have occupational illnesses, and will not have a lot of occupational stress. Despite the fact that the above-mentioned goals are very human, but positive ergonomics advocates claim that the twenty-first century ergonomics should not remain confined to the prevention of occupational diseases, and work accidents only, but should move forward to the promotion of strengths and virtues such as happiness, tranquility, hope, psychological stability, and appreciation. Positive ergonomists see that the promotion of theses virtues and strengths help human beings to overcome the pressures that lead to disturbances of mental health, such as anxiety, depression, despair, and lack of self-esteem. Therefore, positive ergonomics works at two levels: the preventive

http://dx.doi.org/10.5772/67064

level and the therapeutic level. At the preventive level, it strengthens the human psychological adjustment. However, on the therapeutic one, it helps who suffers from maladjustment to strengthen adjustment mechanisms.

Development: During the nineteenth century and before, emotional dimension was an important part in human personality. If researchers were not studying it, they knew it is an important part of the human personality [36–38]. Nevertheless, in the twentieth century especially after the reign of positivism, human scientists did not neglect the emotional dimension only, but even the cognitive and spiritual dimensions on the ground that these dimensions are not measurable and quantifiable. In the second half of the twentieth century, and after the rise of the cognitive, humanistic, and religious schools in human sciences, researchers started shedding light on subjects such as cognition, emotions, and spirit. The result was, many research papers published in many fields, including positive ergonomics.

Positive design: According to Desmet and Pohlmeyer [39], positive design is an umbrella term for many forms of design (design for experience, design for human capabilities, design for socially constructive behavior, design for social innovation, and design for well-being) that affect subjective well-being of individuals and communities. Positive design seeks to achieve the following two goals: to increase the subjective well-being of individuals, and to raise an enduring appreciation of their lives. In some cases, it may not be possible to achieve simultaneously both goals but at least to a certain degree. The above-mentioned authors state that the three ingredients of positive design are design for pleasure, design for personal significance, and design for virtue. If these components are met in a particular design, it is called "a flourishing design", which means a design that helps individuals to flourish or using Maslow's words: "to become actualized in what he is potentially" [40].

The key components of positive design are happiness and comfort.

Happiness: Human beings have been interested in happiness for years. Various approaches have been given to explain what happiness means [41]. The most famous ones are:

- The Hedonic approach to happiness: Happiness is a matter of personal feeling. The more pleasant experiences the individual comes across, the more happiness he has. Everybody feels it in his/ her way depending mainly on his/her personality constitution and social uprising. Some key advocates of this approach are Hobbes [42] and Kahneman et al. [43].
- The desire approach to happiness: According to this approach, happiness is dependent on the individual getting what he/ she wants. The more you get of what you want, the happier you are. One of the key advocates of this approach is Davis [44].
- The spiritual approach to happiness: This approach dates back to the middle-age philosophy. According to this approach, happiness is achieved when the individual converts his/ her soul from acquisitiveness and materialism to complete devotion to God. One of the great advocates of this approach is Al-Ghazzali [45]. In his famous book "The Alchemy of Happiness", stresses that happiness requires committing oneself to the teachings of the Islamic religion, and avoid committing sins. According to him, there are four main

constituents of happiness: self-knowledge, knowledge of God, knowledge of this world as it really is, and the knowledge of the next world as it really is.

- The positive psychology approach to happiness: This approach stresses the connection between thinking about happiness and doing what leads to it. One of the representatives of this approach is Seligman. In his 2011 book [46], Flourish: A Visionary New Understanding of Happiness and Well-being, he mentions that the keys to happiness are PERMA that is achieved through: improving (P) positive emotions, (E) engagement, (R) relationships, (M) meaning, and (A) accomplishment.

Comfort: When a person feels in a particular time he/ she is not suffering from pain and hardship he/she may be in comfort. But if he/she is having a certain amount of hardship, he/ she is uncomfortable. Generally speaking, human beings ought to have a certain amount of comfort which can either be big or small. This amount can last for long time or for a short time. Ergonomics from its emergence advocated that it aims to achieve various goals including human comfort. Seeing various definitions especially the ones of the 1970s and 1980s, comfort is one of these aims. The following are examples:

- Ergonomics is an applied area of psychology and engineering which concerns itself with the design of the physical conditions, machines, and other equipment in relation to human capabilities, learning capacities, efficiency, and comfort [47].
- Ergonomics is a branch of psychology concerned with the design of environments and equipment that promote optimum use of human capabilities and optimum efficiency and comfort [48].
- Ergonomics is the application of information about human characteristics, capacities, and limitations to the design of machines, machine-systems, and environments so that people can live and work safely, comfortably, and effectively. The term also designates the profession that deals with such problems [49].
- Ergonomics studies the role of humans in complex systems, the design of equipment and facilities for human use, and the development of environments for comfort and safety [50].

The positive ergonomics components: The major components of positive ergonomics are: emotional and spiritual ergonomics.

Emotional ergonomics: Researchers [51–54], believe that emotions can be classified on basis of their content, into positive such as joy, contentment, interest and love, and negative emotions such as fear, anger, disgust, and shame. Each emotion either positive or negative is associated with a tendency towards a specific type of action that enhances either success or survival in the specific context. Positive emotions work like a reward that strengthen behavior, whereas negative emotions work like a punishment to stop a behavior from occurring [55, 56]. The reader may come across other related concepts such as mood and affect. In order not to confuse these concepts with emotion, each is defined. First, all these concepts refer to feeling human beings experience. Second, each refers to a particular type of feeling as follows:

- Emotions are intense feelings that generally have well-defined subject (human beings or objects).
- Moods are feelings that are less intense as emotions. Besides, they often have no well-defined subject.
- Affects are feelings that are either intense (emotion) or moderate (mood) feelings.

Definition: Emotional ergonomics pays attention to the emotional aspects of the man-machine system. Man, when in the man-machine system, cannot free himself from his emotions even if they tried to do so. Those who consider that a human being could leave aside his emotions while working do not know the truth about man. A healthy man-machine system is one where workers are allowed to express their emotions [57]. Expressing positive emotions will help man-machine system to obtain favorable outcomes; on the other hand, negative emotions will prevent man-machine system from obtaining positive results [58].

Associated with emotional ergonomics, a new concept used in ergonomics, which is hedonomics. According to Hancock et al. [5], hedonomics refers to the scientific study of enjoyable aspects of man-machine system. Since the concept is new, Helander [59] believes that the problem of ergonomists is now to conceptualize hedonomics, to propose theories that can be used for design, and to build appropriate measurement tools.

It is to note that other terms have been given for hedonomics. Blythe et al. [60] used the term "funology". Jordan [24] used "design for pleasure", Nagamachi [61], used "Kansei engineering", and Helander and Khalid [62] used "affective design".

At the early years of ergonomics, ergonomists were more concerned about physical and cognitive aspects of the system. But, after more than 60 years of ergonomic research, and in response to well spread tendency of positiveness in human sciences, they are very enthusiastic to shed light on the affective side of the man-machine system with the aim of making it pleasurable.

Development: Gendron and Feldman-Barrett [63] think that contemporary psychological research on emotions has gone through three clearly identified stages. These are:

1. The golden stage (before behaviorism): In this stage, research on emotions flourished at the hands of researchers like Darwin [36], James [37], Spencer [64], and Cannon [65].
2. The dark stage (behaviorism time): In this stage, researchers stopped research in the emotion subject. In almost 40 years, nothing worthwhile was published. In fact, it was the dominance of the behavioral approach that marginalized the topic of emotions, as it is of little significance [66, 67].
3. The renaissance stage (from 1960s): In this stage, interest returned again to the topic of emotions which was behind the expansion of conceptual and empirical research, providing a number of influential ideas and theories. One of the influential ideas is what Zajonc [68] had mentioned that emotions function independently of cognition and sometimes it dominates it. Some of the significant theories are: Zajonc theory of emotion [69], the

theory of the evolutionary adaptive value of emotions [70], the broaden-and-build theory of positive emotions [71], and three levels of product emotion theory [72].

Emotional design: It is a design that takes into consideration emotions of users while designing or redesigning goods and products. It maximizes the good (positive) emotions and minimizes the bad (negative) one. In some cases, other names (hedonic design, affective design, affective human factors design, human-centered design, empathetic design) are used for emotional design. Researchers have found that products designed according to emotional design induce among consumers positive feelings which in turn increase pleasure and satisfaction. A good example of emotional design is any product (or features of a product) you loved (you love) in your life.

There are a lot of books written about the design, but the books about emotional design are quite new. They are the result of emotion renaissance era mentioned above. Some of these worthy emotional design books are:

- Seductive Interaction Design: Creating Playful, Fun, and Effective User Experiences by: Anderson [73].
- Designing for Emotion by Walter [22].
- Designing Pleasurable Products: an Introduction to New Human Factors by Jordan [24].
- Designing Emotions by Desmet [74].
- Design and Emotion by McDonagh et al. [75].
- Design for Emotion, by van Gorp and Adams [76].

For Norman [77], emotional design has three levels as follows:

3.2.1. First level, the visceral level

Definition: The visceral level is a level of design in which appearance is very important. It is the initial and immediate reaction to any design that is generally coming from strong emotions and not from logic or reason. These immediate reactions whether in the form of likes or dislikes are often learned. When babies are born, they do not have likes or dislikes, but by time, they learn to like some things (food, drink, cloths, etc…), and to dislike other things. Since these likes and dislikes have been learned, they can be changed by learning as well. In the area of marketing, companies these days rely heavily on this type of design to sell the outputs of goods and services provided. Sometimes, when an enterprise feels that a particular commodity is experiencing a recession, they quickly resort to improve the external appearance. It has been found that product's visceral design communicates to consumers' diverse values. Some of which are aesthetic, functional, and ergonomic values [78]. In addition, it has been found that visceral design helps create commercial success [79]. Nanda et al. [80] found that varying the aesthetic design of the product they studied (BlackBerry Pearl) had a significant impact on emotional reaction of subjects. They found that subjects preBfer the original, piano black treatment of the BlackBerry Pearl over the visually treated versions of the smartphone.

Measurement: It is measured by observing immediate reactions of the individual when he/she is put in front of a design.

Design requirements: It requires the knowledge, skills and abilities of designer. When designers design a particular item, it can be liked or disliked by consumers. If positive feedback is to be given to designer, the product physical features (look, sound, feel, taste, and smell) should be in line with consumer characteristics. When these criteria are taken into account, no doubt both designer and consumer will be pleased. But on the other hand, if negative feedback is given, both designer and consumer are frustrated. Why this kind of stressing situation happens? In situations where products are designed away from the customers' physical, cognitive, emotional, and spiritual information, usually these are the results that are commonly attained.

3.2.2. *Second level, the behavioral level*

Definition: The behavioral level is a level of design in which performance not appearance which matters. In the previous (visceral) design, appearance was essential. Here performance of the product is the most important criteria for choosing that product. If a pen is not writing, it is not a good product. Similarly, if a shaver cannot shave smoothly, it is not a good product. Products can have many other uses, but what is important, is the use for which the product was basically designed. As an example is the chair. Chairs are normally designed for sitting. But, in some cases, they are used to reach the top of a tall shelf. In some other cases, they are used as a hanger to hang the coat in office work, or as a lock to keep the door of a room open, or as a weapon in cases of fight. The design of cup holders in cars is a good example of behavioral design. May be you know that the German car company (BMW) refused to incorporate it in its cars under the pretext that car are for driving, not for drinking. However, when BMW car sales went down, they had (around 2011) to incorporate it.

Measurement: Behavioral design can be measured using well-known methods such as observation, questionnaires, interviews, focus groups, etc.

Design requirements: To accomplish behavioral design, designers should be able to discover the consumers' needs associated with products' functions, usability, understandability, and physical feel. To achieve this goal, designers are advised to carry out field observation and to interview focus groups individuals, so that needs are clearly seen.

3.2.3. *Third level, the reflective level*

Definition: The reflective level is a level of design in which emotions not performance appearance which matter. It can be called emotional design. In this type of design, emotions play a great role. It is based on the idea that emotions function independently of cognition. It is the design that makes the customer to build a positive association with the product at hand. This positive association boosts the sales level. In marketing and advertising sciences, reflective design is becoming very essential. It has been found that an effective reflective design enhances the emotional connection consumers have with brands [81], ad likeability [82], and recall [83].

It is this design that makes a product very liked in one area/ culture, and disliked in another area/ culture.

Measurement: Similar to behavioral design, reflective design can be measured using well-known methods such as observation, questionnaires, interviews, focus groups, etc.

Design requirements: Designers who want to produce design at this level should not base their designs on their emotions themselves. Despite the fact that customers' emotions are diversified and varied, they ought to understand them and design goods and products accordingly.

Wherever you go, you are plagued with so many products with very different designs. These designs can be good or bad designs. It depends on the user judgments. If a particular design pleases the user, it is a good design. Otherwise it is not. It has been believed that ergonomically designed products are automatically pleasurable products [84]. However, from the emotional design point of view, ergonomically designed products, may be efficient ones, but they may not be pleasing the intended users. In the area of multimedia learning material, it has been found that if learning evokes positive emotions among learners, the learning process will be facilitated [85].

Spiritual ergonomics: As long as all workers or at least the majority of workers believe in Allah/ God/ or other universal spirit, they bring with them to man-machine system, their belief and faith and they will be very pleased if their spiritual needs are satisfied in their work. Workers who are denied spiritual needs are not motivated to do the work and may be subjected to work stress, accidents, and other diseases. In the first two or three decades of ergonomics, attention was paid to workers bodies and cognitions; however, in the recent decades attention was shifted to emotional and spiritual sides of workers.

Definition: Spiritual ergonomics pays a lot of attention to the spiritual side of the man in the man-machine system. It is based on the idea that spirit is a key factor which determines the employee's health and success in the man-machine system, no matter what he/she is doing in that system. Spiritualism is important to individuals for various reasons. Spiritual people are generally satisfied with their lives in comparison to their counterparts who are not spiritual. Some are:

- Spiritual people give meaning to life.
- Spiritual people do not lose hope.
- Spiritual people do not suffer or suffer little stress.
- Spiritual people are covered with happiness and contentment.
- Spiritual people self-actualize.

Development: Spiritual ergonomics is a part of the spiritual movement which has spread in the world especially in the seventies, for several reasons. First is the perceived failure of Christianity. Christianity over the years suffered a lot of criticism. Christianity is related to colonialism. Also it is related to and supported for dictatorship and totalitarian regimes.

Second is the failure of secular humanism to provide spiritual and ethical guidance for the future. In addition, researchers like Cacioppe [86] argue that modern world is afflicted by social, economic, and environmental problems that are caused mainly by disabling the spiritual aspects of human beings. The absence of these spiritual aspects led to the absence of love, compassion, altruism, and helping behavior. The search for these virtues is essentially a spiritual journey.

Spiritual design: Initially, the researcher notes that he could not find past studies and literature on this design. Research in this area has been delayed too much from the rest of the designs that have been referred to. However, it is a design that takes into consideration spiritual needs of users while designing or redesigning goods and products. Spiritual needs differ depending on the religious education the individual gets upon socialization. What a Christian person gets is to a great extent different from what a Muslim or Jewish or Buddhist gets. As stated by Highfield and Cason [87], the spiritual needs that patients seek to satisfy are the following four:

- The need for meaning and purpose.
- The need to give love.
- The need to receive love.
- The need for forgiveness, creativity, and hope.

On the other hand, in Islam, the very important spiritual needs are:

- Needs related to acts of worship such as prayer, and fasting.
- Needs related to morality as good dealing with people and honesty.
- Needs related to the relationship between the individual and Allah (God) as fear, hope, humility, sincerity, and kindness.

Measurement: Researchers have done a lot of work to measure spiritual design. The result is a number of tools that can be used. But it has to be mentioned that the choice of a particular tool is subject to a lot of considerations foremost, religion owed by the individual and the culture in which he/she resides. Some of these tools are observation, interviews, questionnaires, and other scales. For some of these tools, you can study:

1. The Spirituality Index of Well-Being by Daaleman and Frey [88].
2. The Spirituality and Spiritual Care Rating Scale by McSherry et al. [89].
3. The Spiritual Care Competence Scale by van Leeuwen et al. [90].

Design requirements: Goods and products are not randomly designed by engineers. If they want them to be desirable commodities, and customers to be happy with them, they must include what states that the spiritual aspect of customers has been taken into account. Spiritual values of people should be known to designers and engineers, otherwise harmony with spiritual values of individuals may not be possible.

3.3. What is the relationship of positive ergonomics with occupational health?

Occupational health refers to a series of actions that aim at protecting workers and reducing the risk of equipment, and machinery and attempting to prevent work accidents, injuries and illnesses, and providing safe work environment. Occupational health intervenes in all areas of life.

In 1957, the first Joint ILO/WHO Committee of Occupational Health assembled, and identified areas of interest in occupational health in the following: (1) the maintenance and promotion of workers' health and working capacity; (2) the improvement of working environment and work to become conducive to safety and health; (3) the development of work organizations and working cultures in a direction which supports health and safety at work; and (4) the promotion of a positive social climate to smooth operations and enhance productivity of the undertakings [91].

Alli [92] believes that occupational safety and health is "the science of the anticipation, recognition, evaluation, and control of hazards arising in or from the workplace that could impair the health and well-being of workers".

But how occupational safety tried to implement industrial safety at work environment? When considering the legacy of occupational health, we find that two approaches were adopted throughout of its history. These are: professional harmonization and ergonomics approaches.

The professional harmonization approach is concerned with minimizing loss by aiding in the preservation and protection of both human and other physical assets in the workplace. It primarily involves monitoring the workplace and advising employers or management on the best ways to prevent and minimize losses [93]. In order for this approach to be successful, authorities must do a series of actions including tough selection and training of workers. Workers are not selected based on fitting the direct requirement of work only, but based on fitting indirect requirements as well. Other features such as accident proneness [94], various obsessions [95], mental toughness [96], etc., are important issues for the success in the twentieth century.

The ergonomic approach is concerned with implementing ergonomics principles at work. Authorities are requested to make sure that ergonomics principles are applied in the work area [97]. When applied, occupational health will improve [98, 99]. Here, it is to mention that designing or redesigning tools, machines, equipment, workplaces, and work environment to fit the physical parts of the workers is not enough. Data concerning other aspects of the worker, for example, cognitive capacities, emotional skills, and spiritual aptitudes should also be collected to be used in designing or redesigning work. This new orientation of ergonomics takes into account the physical, mental, emotional, and spiritual aspects of the human factor. It is based on the idea that the workers can be happy at work only if they feel that the needs of these different aspects are saturated. This is what the new approach to occupational health, aspires to achieve.

To be sure about the effectiveness of twenty-first century occupational health, the two above-mentioned approaches (professional harmonization and ergonomics) are to be sequentially used.

4. Conclusions

The official age of ergonomics is now about 66 years. In this long life, ergonomics has witnessed many developments. Some of them have been on the level of subject matter, for example, cognitive ergonomics in the 1960s, organizational ergonomics in the 1970s, positive ergonomics in the 1980s, emotional ergonomics in 1990s and spiritual ergonomics in the new millennium. Some other developments have been on the methodological level, for example, the use of qualitative methods in the 1980s and mixed methods in the 1990s.

In this chapter, we shed light on subject matter developments where we discussed positive ergonomics with its subsequent design, i.e., the positive design. Then emotional ergonomics with its consequent emotional design were debated. At the end, the newly born development, i.e., the spiritual ergonomics with spiritual design were discussed. Our hope is that future research will ponder greatly into these developments. Time in which these important human personality aspects (emotions and spiritual aspects) have been excluded or left out has gone and may never return.

If ergonomics takes these issues into account, it will not only succeed, but it will spread widely among the people of the earth from different orientations and religions.

Author details

Mohamed Mokdad* and Tawfik Abdel-Moniem

*Address all correspondence to: mokdad@hotmail.com

University of Bahrain, Sakhir, Bahrain

References

[1] International Ergonomics Association (IEA). Definition and Domains of Ergonomics [Internet]. 2016. Available from: http://www.iea.cc/about/index.html [Accessed: 2016-07-12].

[2] Hart, P.M. Predicting Employee Life Satisfaction: A Coherent Model of Personality, Work and Nonwork Experiences, and Domain Satisfactions. Journal of Applied Psychology, 1999;**84**:564–584. DOI:10.1037/0021-9010.84.4.564.

[3] Helander, M.G., Khalid, H. & Tham, M.P. (Eds.). Affective Human Factors Design. London: ASEAN Academic Press; 2001. DOI: 10.1177/154193120204601209.

[4] Martino, O.I. & Morris, N. Positive Ergonomics: Improving Mood Before the Working Day Begins. In: P.T. McCabe (Ed.). Contemporary Ergonomics. New York: CRC Press; 2004, pp. 425–429. DOI: 10.1201/9780203494172.ch77.

[5] Hancock, P.A., Pepe, A.A. & Murphy, L.I. Hedonomics: The Power of Positive and Pleasurable Ergonomics. Ergonomics in Design. 2005;**13**(1):8–14. DOI: 10.1177/106480460501300104.

[6] Seligman, M.E.P. & Csikszentmihalyi, M. Positive Psychology: An Introduction. American Psychologist. 2000;**55**:5–14. DOI: 10.1037/0003-066X.55.1.5.

[7] Diener, E. & Seligman, M.E.P. Beyond Money: Toward an Economy of Well-Being. Psychological Science in the Public Interest. 2004;**5**:1–13. DOI: 10.1111/j.0963-7214.2004.00501001.x.

[8] Huppert, F., Baylis, N. & Keverne, B. The Science of Well-Being. Oxford, England: Oxford University Press; 2006. ISBN: 0198567529.

[9] Licht, D.M., Polzella, D.J. & Boff K.R. Human factors, ergonomics and human factors engineering: An analysis of definitions. Crew System Ergonomics Information Analysis Center; 1989. DOI: 10.13140/RG.2.1.4367.3365.

[10] Wogalter, M.S., Hancock, P.A. & Dempsey, P.G. On the Description and Definition of Human Factors/Ergonomics. In: Human Factors and Ergonomics Society 42nd Annual Meeting; October 5–9, Chicago, Illinois; 1998. pp. 671–674. DOI: 10.1177/154193129804201001.

[11] Dempsey, P.G., Wogalter, M.S. & Hancock, P.A. Defining ergonomics/human factors. In: W. Karwowski (Ed.). International Encyclopedia of Ergonomics and Human Factors. 2nd ed. London: Taylor and Francis; 2006. pp. 32–35. DOI: 10.1201/9780849375477.ch7.

[12] Human Factors and Ergonomic Society (HFES). Definitions of human factors and ergonomics [Internet]. 2012. Available from: http://hfes.org/web/EducationalResources/HFEdefinitionmain.html#govaagencies [Accessed: 2016-07-13].

[13] Chapanis, A.R. The Search for Relevance in Applied Research. In: W.T. Singleton, J.G. Fox & D.C. Whitfield (Eds.). Measurement of Man at Work. 1st ed. London: Taylor & Francis; 1971. pp. 1–14.

[14] Targett, D. Quantitative Methods. 1st ed. Edinburgh: Edinburgh Business School, Heriot-Watt University; 2001.

[15] Hancock, P.A. & Szalma, J.L. On the Relevance of Qualitative Methods for Ergonomics. Theoretical Issues in Ergonomics Sciences. 2004;**5**(6):499–506. DOI: 10.1080/14639220412331303391.

[16] Hignett, S. & Wilson, J.R. The Role for Qualitative Methodology in Ergonomics: A Case Study to Explore Theoretical Issues. Theoretical Issues in Ergonomics Science. 2004;**5**(6):473–493. DOI: 10.1080/14639220412331303382.

[17] Pandve, H.T. Qualitative Research in Ergonomics: An Added Advantage. Journal of Ergonomics (Open access). 2016;**6**:2. DOI: 10.4172/2165-7556.1000e150.

[18] Falzon, P., Gaines, B.R. & Monk, A.F. Cognitive Ergonomics: Understanding, Learning and Designing Human-Computer Interaction. London: Academic Press Ltd.; 1990. ISBN: 978-0122482908.

[19] Hoc, J.M. Cognitive Ergonomics: A Multidisciplinary Venture. Ergonomics. 2008;**51** (1):71–75. DOI: 10.1080/00140130701801124.

[20] Ryff, C.D. & Singer, B. The Contours of Positive Human Health. Psychological Inquiry. 1998;**9**(1):01–28. DOI: 10.1207/s15327965pli0901_1.

[21] Coelho, D.A. A Growing Concept of Ergonomics Including Comfort, Pleasure and Cognitive Engineering: An Engineering Design Perspective [thesis]. Covilhã, Portugal: Department of Electromechanical Engineering, School of Engineering Sciences, Universidade da Beira Interior, Covilhã, Portugal; 2002. DOI: 10.13140/2.1.3690.2403.

[22] Walter, A. Designing for Emotion. 1st ed. New York: Jeffrey Zeldman; 2011. ISBN: 978-1937557003.

[23] Tzou, R.C. & Lu, H.P. Exploring the Emotional, Aesthetic, and Ergonomic Facets of Innovative Product on Fashion Technology Acceptance Model. Behaviour & Information Technology. 2009;**28**(4):311–322. DOI: 10.1080/01449290701763454.

[24] Jordan, P.W. Designing Pleasurable Products: An Introduction to New Human Factors. 1st ed. London: Taylor and Francis; 2000. DOI: 10.1162/desi.2002.18.1.87.

[25] Pimentel, R. Emotional Ergonomics: The Price of Making a Living Shouldn't Be Your Life [Internet]. 2000. Available from: http://www.miltwright.com/articles/emotionalergonomics_price.pdf [Accessed: 2016-07-13].

[26] Delaney, C. The Spirituality Scale: Development and Psychometric Testing of a Holistic Instrument to Assess the Human Spiritual Dimension. Journal of Holistic Nursing. 2005;**23**(2):145–167. DOI: 10.1177/0898010105276180.

[27] Petchsawanga, P. & Duchon, D. Workplace Spirituality, Meditation, and Work Performance. Journal of Management, Spirituality & Religion. 2012;**9**(2):189–208. DOI: 10.1080/14766086.2012.688623.

[28] Wihlidal, L.M. & Kumar, S. An Injury Profile of Practicing Diagnostic Sonographers in Alberta. International Journal of Industrial Ergonomics. 1997;**19**:205–126. DOI: 10.1016/S0169-8141(95)00107-7.

[29] Winzeler, S. & Rosenstein, B. Orthopedic Problems of the Upper Extremity: Assessment and Diagnosis. American Association of Occupational Health Nurses. 1997;**45**(4):188–202. PMID:9155269.

[30] OSHA. Solutions for the Prevention of Musculoskeletal Injuries in Foundries [Internet]. 2012. Available from: https://www.osha.gov/Publications/osha3465.pdf. [Accessed: 2016-07-13].

[31] Anton, D. & Weeks, D.L. Prevalence of work-Related Musculoskeletal Symptoms Among Grocery Workers. International Journal of Industrial Ergonomics. 2016;**54**:139–145. DOI: 10.1016/j.ergon.2016.05.006.

[32] Friedman, L., Krupczak, C., Brandt-Rauf, S. & Forst, L. Occupational Amputations in Illinois 2000–2007: BLS vs. Data Linkage of Trauma Registry, Hospital Discharge, Workers Compensation Databases and OSHA Citations. Injury. 2013;**44**(5):667–673. DOI: 10.1016/j.injury.2012.01.007.

[33] Tompa, E., Dolinschi, R., de Oliveira, C. & Irvin, E. A Systematic Review of OHS Interventions with Economic Evaluations, Vols. 1 and 2. 1st ed. Toronto, Ontario, Canada: Institute for Work & Health; 2007. DOI: 10.1097/JOM.0b013e3181b34f60.

[34] Sinha, V.C., Saxena, J.K. & Gupta, A. Business Statistics. 1st ed. New Delhi: SBPD Publications; 2015. ISBN: 9382883738.

[35] Gable, S.L. & Haidt, J. What (and Why) Is Positive Psychology? Review of General Psychology. 2005;**9**(2):103–110. DOI: 10.1037/1089-2680.9.2.103. 103.

[36] Darwin, C. The Expression of the Emotions in Man and Animals. London: John Murray; 1872. pp. 234–263. DOI: 10.1007/978-3-531-93439-6_11.

[37] James, W. What is an Emotion? Mind. 1884;**9**:188–205. DOI: 10.1093/mind/os-IX.34.188.

[38] Wundt, W. Grundirss der Psychologie (Outlines of Psychology). 1st ed. Leipzig: Engelmann; 1896.

[39] Desmet, P.M.A. & Pohlmeyer, A.E. Positive Design: An Introduction to Design for Subjective Well-Being. International Journal of Design. 2013;**7**(3):5–19. ISSN: 1994-036X (online); 1991-3761 (print).

[40] Maslow, A.H. Towards a Psychology of Being. 1st ed. Princeton: D. Van Nostrand Company; 1962. DOI: 10.1037/10793-000.

[41] Bremner, R.H. Theories of Happiness: On the Origins of Happiness and Our Contemporary Conception [thesis]. Bonn: Faculty of Philosophy, Rheinische Friedrich Wilhelm University, Bonn; 2011. hss.ulb.uni-bonn.de/2011/2597/2597.pdf.

[42] Hobbes, T. Leviathan. 1st ed. Indianapolis, Cambridge: Hackett Publishing Company; 1994. ISBN: 978-0-87220-178-1.

[43] Kahneman, D., Diener, E. & Schwarz, N. Well-Being: The Foundations of Hedonic Psychology. New York: Russell Sage Found; 1999. DOI: 10.1.1.121.6695.

[44] Davis, W. A Theory of Happiness. American Philosophical Quarterly. 1981;**18**:111–120.

[45] **Al**-Ghazzali, I.M. The Alchemy of Happiness. 1st ed. USA: SIME Journal; 2004. DOI: 10.1017/S1356186300003771.

[46] Seligman, M.E.P. Flourish. 1st ed. New York: Free Press; 2011. ISBN 9781439190760.

[47] Wolman, B.B. (Ed.). Dictionary of Behavioral Science. 1st ed. New York: Van Nostrand Reinhold; 1973. ISBN-13: 978-0127624556.

[48] Goldenson, R.M. (Ed.). Longman Dictionary of Psychology and Psychiatry. 1st ed. White Plains, NY: Longman Incorporated; 1984. DOI: 10.1108/09504120710737987.

[49] Chapan is A.R. Human Engineering. In: Holland DT, editor. Encyclopedia Americana. 1st ed. New York: Grolier; 1986. p. 549b.

[50] Salvendy, G. (Ed.). Handbook of Human Factors. 1st ed. New York: John Wiley & Sons; 1987. DOI: 10.1002/(SICI)1520-6564(199922)9:3<321::AID-HFM9>3.0.

[51] Frijda, N.H. The Emotions. 1st ed. Cambridge, MA: Cambridge University Press; 1986.

[52] Lazarus, R.S. Emotion and Adaptation. 1st ed. New York: Oxford University Press; 1991. ISBN: 9780195069945.

[53] Levenson, R.W. Human Emotion: A Functional View. In: P. Ekman & R.J. Davidson (Eds.). The Nature of Emotion: Fundamental Questions. 1st ed. New York: Oxford University Press; 1994. pp. 123–126. ISBN: 0195089448, 9780195089448.

[54] Oatley, K. & Jenkins, M. Understanding Emotions. 1st ed. Cambridge, MA: Blackwell; 1996. ISBN-13: 978-1557864956.

[55] Larsen, R.J. & Ketelaar, T. Extraversion, Neuroticism, and Susceptibility to Positive and Negative Mood Induction procedures. Personality and Individual Differences. 1989;**10**:1221–1228. DOI: 10.1016/0191-8869(89)90233-X.

[56] Spoor, J.R. & Kelly, J.R. The Evolutionary Significance of Affect in Groups: Communication and Group bonding. Group Processes and Intergroup Relations. 2004;**7**:398–412. DOI: 10.1177/1368430204046145.

[57] Staw, B.M, Sutton, R.I & Pelled, L.H. Employee Positive Emotion and Favorable Outcomes at the Work-place. Organization Science. 1994;**5**:51–71. DOI: 10.1287/orsc.5.1.51.

[58] Canaff, A.L. & Wright, W. High Anxiety: Counselling the Job- Insecure Client. Journal of Employment Counselling. 2004;**41**(1):2–10. DOI: 10.1002/j.2161-1920.2004.tb00872.x.

[59] Helander, M.G. Hedonomics–Affective Human Factors Design. Proceedings of the Human Factors and Ergonomics Society Annual Meeting. 2002;**46**(12):978–982. DOI: 10.1080/00140130310001610810.

[60] Blythe M.A., Overbeeke K, Monk A.F., Wright, P.C., editors. Funology: from usability to enjoyment. Springer Science & Business Media; 2004 Oct 13. DOI: 10.1.1.372.6229

[61] Nagamachi, M. Kansei Engineering: A New Ergonomic Consumer-Oriented Technology for Product Development. International Journal of Industrial Ergonomics. 1995;**15**:3–11. DOI: 10.1016/0169-8141(94)00052-5.

[62] Helander, M.G. & Khalid, H.M. Underlying Theories of Hedonomics for Affective and Pleasurable Design. Proceedings of the Human Factors and Ergonomics Society Annual Meeting. 2005;**49**(18):1691–1695. DOI: 10.1177/154193120504901803.

[63] Gendron, M. & Feldman-Barrett, L. Reconstructing the Past: A Century of Ideas about Emotion in Psychology. Emotion Review. 2009;1(4):316–339. DOI: 10.1177/1754073909338877.

[64] Spencer, H. The Principles of Psychology. 1st ed. London: Williams and Norgate; 1870.

[65] Cannon, W.B. The James-Lange Theory of Emotion: A Critical Examination and an Alternative theory. American Journal of Psychology. 1927;**39**:106–124. DOI: 10.2307/1415404.

[66] Clark, M.S. (ed.). Emotion and Social Behavior. Review of Personality and Social Psychology, Vol. 14. New York: Sage; 1992.

[67] Watson, D. & Tellegen, A. Towards a Consensual Structure of Mood. Psychological Bulletin. 1985;**98**:219–235. DOI: 10.1037/0033-2909.98.

[68] Zajonc, R.B. Feeling and Thinking: Preferences Need no Inferences. American Psychologist. 1980;**35**(2):151–175. DOI: 10.1037/0003-066X.35.2.151.

[69] Zajonc, R.B. Emotion and Facial Efference: a Theory Reclaimed. Science. 1985;**228**:15–21. DOI: 10.1126/science.3883492.

[70] Rolls, E.T. A Theory of Emotion, and its Application to Understanding the Neural Basis of Emotion. Cognition & Emotion. 1990;**4**:161–190. DOI: 10.1080/02699939008410795.

[71] Fredrickson, B.L. The Broaden-and-Build Theory of Positive Emotions. Philosophical Transactions of the Royal Society B Biological Sciences. 2004;**359**(1449):1367–1378. DOI: 10.1098/rstb.2004.1512.

[72] Desmet, P.M.A. Three Levels of Product Emotion. In: International conference on Kansei engineering and emotion research 2010 (KEER 2010); March 2–4 2010; Paris. 2010.

[73] Anderson, S.P. Seductive Interaction Design: Creating Playful, Fun, and Effective User Experiences. Berkeley, CA, USA: New Riders; 2011. ISBN-13: 978-0321725523.

[74] Desmet, P.M.A. Designing Emotions. 1st ed. The Netherlands: TU Delft, Delft; 2002. ISBN: 90-9015877-4.

[75] McDonagh, D., Hekkert, P., van Erp, J. & Gyi, D. Design and Emotion. 1st ed. USA: CRC Press; 2004. DOI: 10.1080/09544820902915300.

[76] van Gorp, T. & Adams, E. Design for Emotion. 1st ed. London: Elsevier; 2012. DOI: 10.1016/B978-0-12-386531-1.00001-6.

[77] Norman, D.A. Emotion and Design: Attractive Things Work Better. Interactions Magazine. 2002;**4**(4):36–42. DOI: 10.1145/543434.543435.

[78] Schoormans, J.P.L. & Robben, H.S.J. The Effect of New Package Design on Product Attention, Categorization and Evaluation. Journal of Economic Psychology. 1997;**18**:271–287. DOI: 10.1016/S0167-4870(97)00008-1.

[79] Yamamoto, M. & Lambert, D.R. The Impact of Product Aesthetics on the Evaluation of Industrial Products. Journal of Product Innovation Management. 1994;**11**(4):309–324. DOI: 10.1111/1540-5885.1140309.

[80] Nanda, P., Bos, J., Kramer, K.L., Hay, C. & Ignacz, J. Effect of Smartphone Aesthetic Design on Users' Emotional Reaction: An Empirical Study. The TQM Journal. 2008;**20**(4):348–355. DOI: 10.1108/17542730810881339.

[81] Mehta, A. & Purvis, S.C. Reconsidering Recall and Emotion in Advertising. Journal of Advertising Research. 2006;**46**(1):49–56. DOI: 10.2501/S0021849906060065.

[82] McDuff, D., El Kaliouby, R., Kodra, E. & Picard, R. Measuring Voter's Candidate Preference Based on Affective Responses to Election Debates. In: Humaine Association Conference on Affective Computing and Intelligent Interaction (ACII); September 2013; NewYork, USA: Institute of Electrical and Electronics Engineers; 2013. pp. 369–374. DOI: 10.1109/ACII.2013.67.

[83] Hazlett, R.L. & Hazlett, S.Y. Emotional Response to Television Commercials: Facial EMG Vs. Self-Report. Journal of Advertising Research. 1999;**39**(2):7–23.

[84] Iqbal, M., Iqbal, S.A., Mustafizur Rahman, A.N. & Samsuzzoha, A.H.M. Ergonomics and Design. In: Proceedings of the 2011 International Conference on Industrial Engineering and Operations Management, Kuala Lumpur, Malaysia; January 22–24, 2011; 2011.

[85] Heidig, S. Müller, J. & Reichelt, M. Emotional Design in Multimedia Learning: Differentiation on Relevant Design Features and their Effects on Emotions and Learning. Computers in Human Behavior. 2015;**44**:81–95. DOI: 10.1016/j.chb.2014.11.009.

[86] Cacioppe, R. Creating Spirit at Work: Re-visioning Organization Development and Leadership-Part I. Leadership and Organization Development Journal. 2000;**21**(2):48–54. DOI: 10.1108/01437730010318200.

[87] Highfield, M.F. & Cason, C. Spiritual Needs of Patients: Are They Recognized? Cancer Nursing. 1983;**6**(3):187–192. PMID: 6190552.

[88] Daaleman, T.P, Frey, B.B. The Spirituality Index of Well-Being: A New Instrument for Health-Related Quality-of-Life Research. Annals of Family Medicine. 2004;**2**:499–503. PMID: 15506588.

[89] McSherry, W., Draper P. & Kendrick D. The Construct Validity of a Rating Scale Designed to Assess Spirituality and Spiritual Care. International Journal of Nursing Studies. 2002;**13**(7):723–734. PMID: 12231029.

[90] van Leeuwen, R., Tiesinga, L.J., Middel, B., Post, D. & Jochemsen, H. The Validity and Reliability of an Instrument to Assess Nursing Competencies in Spiritual Care. Journal of Clinical Nursing. 2009;**18**(20):2857–2869. DOI: 10.1111/j.1365-2702.2008.02594.x.

[91] Board, 20. 1957 Report of the Joint ILO/WHO Committee on Occupational Health. 1st ed. Geneva: World Health Organization; 1957.

[92] Alli, B.O. Fundamental Principles of Occupational Health and Safety. 1st ed. Geneva: International Labour Office; 2008. ISBN: 9221204545.

[93] Friend, M.A. & Kohn, J.P. Fundamentals of Occupational Safety and Health. 6th ed. London: Brenan Press; 2014. ISBN-13: 978-0-86587-171-7.

[94] Visser, E., Pijl, Y.J., Stolk, R.P., Neeleman, J., Rosmalen, J.G.M. Accident Proneness, Does it Exist? A Review and Meta-Analysis. Accident Analysis and Prevention. 2007;**39**:556–564. DOI: 10.1016/j.aap.2006.09.012.

[95] Rachman, S. A Cognitive Theory of Obsessions. Behaviour Research and Therapy. 1997;**35**(9):793–802. PMID: 9299799.

[96] Connaughton, D., Wadey, R., Hanton, S. & Jones, G. The Development and Maintenance of Mental Toughness: Perceptions of Elite Performers. Journal of Sport Sciences. 2008;**26**:83–95. DOI: 10.1080/02640410701310958.

[97] OSHA. Ergonomics: The Study of Work. 1st ed. USA: U.S. Department of Labor Occupational Safety and Health Administration; 2000.

[98] Burri Jr., G.J. & Helander, M.G. A Field Study of Productivity Improvement in the Manufacturing of Circuit Boards. International Journal of Industrial Ergonomics. 1991;**7**:207–215. DOI: 10.1016/0169-8141(91)90004-6.

[99] Corlett, E.N. Are You Sitting Comfortably? International Journal of Industrial Ergonomics. 1999;**24**:7–12. DOI: 10.1016/S0169-8141(98)00083-3.

Chapter 2

Understanding the Stakeholders as a Success Factor for Effective Occupational Health Care

Ari-Matti Auvinen

Additional information is available at the end of the chapter

http://dx.doi.org/10.5772/66479

Abstract

Effective occupational health care at the workplace requires collaboration, partnerships and alliances with internal, interface and external stakeholders. Essential steps for solid work with various stakeholders are identification of key stakeholders, systematic analysis of their views and positions, and development of stakeholder participation and involvement. Stakeholder analysis aims to evaluate and understand stakeholders from the perspective of an organization. Stakeholder analysis starts with identifying and classifying the key stakeholders. After their identification, questions are asked about their position, interest, influence, inter-relations, networks and other characteristics of stakeholders, with reference to their past and present positions, and future potential. The results are presented as stakeholder maps as well as by the power-interest matrix of the stakeholders. Stakeholder analysis serves an organization and its various actors as a guideline in identifying, planning and implementing strategies for managing stakeholder relationships and utilizing the full potential of various stakeholders in developing occupational health care.

Keywords: stakeholders, stakeholder analysis, power-interest matrix, stakeholder strategy

1. Introduction

In the modern operating environment of work, successful occupational health care requires collaboration, partnerships and alliances between various actors. The various actors with clear interests ('stakes') in the work and operations within a workplace are called stakeholders. The changing nature of workplaces is accelerating the needs for collaboration between various stakeholders.

Some decades ago, the focus of occupational health care was either on a single illness or risk factor, or on changing a particular lifestyle habit or behaviour of individual employees. In countries like Finland, the development has clearly been towards a comprehensive approach in occupational health care, and thus the focus has been shifted towards active promotion of the work ability of individuals with an emphasis on structural developments [1]. In general, since the 1990s in many industrialized countries, occupational health care has been understood as more holistic and integrative by its nature [2].

Occupational health care can be understood today as the combined efforts of employers, employees and society to improve the health and well-being of people at work. This conceptual development has also meant that the number of actors in occupational health care has been increasing. Consequently, new types of collaboration between employees, employers and other actors in the field are required [1, 3]. The altering operational environment also requires new skills, such as networking skills, which are becoming essential for an effective occupational health care work [4].

Various stakeholders of occupational health care (e.g. employees, employers, shareholders and occupational healthcare providers) also have different key interests [5]. If their different interests are recognized and analysed, the operating environment of occupational health care can be improved. However, as the "Healthy workplaces: a model for action" by the WHO emphasizes, the various stakeholders in a workplace must work together in a collaborative manner [6].

The classic definition of a stakeholder according to Freeman is "an organization… [or] any group or individual who can affect or be affected by the achievement of the organization's objectives" [7]. While Freeman's groundbreaking book "Strategic Management: A Stakeholder Approach" in 1984 started the wider discussion and elaboration of stakeholders and their importance, an earlier concept of stakeholders had already emerged in the 1960s. In 1963, academics at the Stanford Research Institute stated that a firm also needs to be responsible – in addition to shareholders – to a number of stakeholders without whose support the organization would cease to exist [5]. Some scholars have even proposed that the roots of stakeholder thinking dates as far back as the 1930s [8].

The number of published titles in academic and professional literature about stakeholder management has grown rapidly since the 1980s. The main body of the stakeholder management literature still relates to the corporate environment, but the public sector [9] and the third sector [10] are becoming increasingly interested in the applications of stakeholder management as an element in their strategic management.

The concept of stakeholders and the potential for both convergent and competing issues in the corporate world have often been displayed through the consideration of corporate social responsibility and corporate ethics [10–12]. Some researchers argue that the fundamental concepts of stakeholder, stakeholder model, stakeholder management and stakeholder theory are explained by various authors in different ways and are supported or critiqued with diverse and often contradictory evidence and arguments [13].

The scientific literature concerning stakeholders in occupational health care is evolving, but still scarce, and there is an obvious lack of published research in this area [14]. The theme of stakeholders has been discussed in the context of national healthcare systems and healthcare organizations (see e.g. [15–18]). The importance of stakeholders and stakeholder positions for hospitals has also been discussed in the research literature (see e.g. [19, 20]). The importance of stakeholders in occupational health services has mainly been discussed in the context of the need for multiple stakeholder collaboration (see e.g. [21]), but a generic, structured approach to understand stakeholders and their positions in occupational health care is still covered only by few authors in the research literature.

Understanding the stakeholders is essential also in the altering environment of value creation in occupational health care. In modern approach to occupational health care, the individuals (employees) of an organization are essential for the value creation. As the most important determinant of an individual's health is his or her own health behaviour, the different types of service can support an individual in co-creating better health [22]. Co-creation implies meaningful engagements of interaction, activities and exchange between collaborators [22]. The co-creation approach is valuable, as it emphasizes the critical role and the involvement of the users in the value creation, but also the various encounters with different actors.

The critical involvement of the users is described with the term 'value co-creation' and the value provision as a collaborative action between various actors and players is described with the term 'value co-production'. Joint actions of various entities can provide novel opportunities and avenues for various users. Value co-production as such is not a recent innovation; the recent innovation is to organize the value co-production systematically [23]. Effective occupational health care should utilize both modalities, as it needs both the strong involvement of various actors into value creation as well as well-organized collaboration between various actors to enable comprehensive service provision for the users of occupational health care [24].

The implications of co-production are also visible to occupational health care and its value creation. The value creation can take place in a more synchronous, less sequential manner by various actors. However, it also provides new opportunities to provide a comprehensive service provision for the users of occupational health care [23].

The altering operating environment of occupational health care, its novel challenges and new opportunities in value creation urge also the more in-depth knowledge and analysis of the various stakeholders.

2. Stakeholder analysis as an important tool

The aim of stakeholder analysis is to evaluate and understand stakeholders from the perspective of an organization, or to determine their relevance to a project or policy [16]. In the undertaking of stakeholder analysis, various questions are asked about the position, interest, influence, inter-relations, networks and other characteristics of stakeholders, with reference to their past and present positions, and future potential [25].

More specifically, stakeholder analysis is an approach, a tool or set of tools for generating knowledge about 'actors' – individuals and organizations – so as to understand their behaviour, intentions, inter-relations and interests. Furthermore, such analysis is beneficial in the assessment of the influence and resources stakeholders bring to bear on the decision-making or the implementation process [26].

Stakeholder analysis is the essential tool to be used in stakeholder management. Stakeholder management is an approach to strategic management of an organization, which emphasizes the crucial role of different stakeholders in the success of the operations of an organization [7, 27]. Stakeholder analysis and stakeholder management are of particular importance to public and non-profit organizations, which have more diverse groups of stakeholders than private for-profit organizations [9].

Although stakeholder analysis is an important building block in stakeholder management, stakeholder analysis itself can make a significant contribution as a research method and as a means of organizational change [28]. Thus stakeholder analysis presents its own, independent and intrinsic value even if an organization is not implementing a thorough stakeholder management approach.

'Stakeholder analysis' is not a clearly defined analysis technique; rather it includes an array of various techniques. Bryson has identified and presented 15 stakeholder identification and analysis techniques [9]. In this article, the stakeholder analysis is described to include the following key stages:

- identification of important stakeholders
- mapping and assessing of stakeholder positions and views
- undertaking a diagnosis of stakeholder positions and views.

Auvinen et al. used a similar approach in mapping and analysing the positions of Finnish stakeholders in workplace health promotion [14].

The first stage is to identify the key stakeholders. In the stakeholder literature, the definition of the 'wide sense of stakeholders' and the 'narrow sense of stakeholders' is an essential element (see e.g. [29, 30]). The 'narrow sense of stakeholders' limits the scope of stakeholders to groups or individuals who can affect the achievement of an organization's objectives or who are affected by the achievement of an organization's objectives; the 'wide sense of stakeholders' widens the scope of stakeholders to identifiable groups or individuals on which the organizations is dependent for its continued survival [29]. In this article, the wide sense of a stakeholder was applied, and thus the stakeholders were defined as all those who have a legitimate interest (either direct or indirect) in occupational health care and its activities.

The following stage is to classify the stakeholders as either primary or secondary stakeholders [9, 31]. According to Clarkson's widely used definition, the primary stakeholder groups are ones without whose continuing participation the corporation cannot survive as a going concern. Furthermore, the secondary stakeholder groups are those who influence or affect, or are influenced or affected by, the organization, but they are not engaged in transactions with the corporation and are not essential for its survival [31].

The third stage in the stakeholder analysis is to assess the positions and views of the stakeholders, and based on this assessment construct a view of their 'relative importance'. As the analysis of the dynamics of the stakeholders is essential, this stage provides the basis for important activity. The literature of stakeholder analysis and management is rich on varieties of classifications of the stakeholders. Some authors classify the stakeholders according to power, legitimacy and urgency (see e.g. [30]) while another author classifies them according to the issue position and importance [9]. For the purposes of this article, the power/interest matrix was used, as it provides a comprehensive and understandable tool to present positions of the stakeholders (see also [32]).

3. Identification of key stakeholders

3.1. Classification of various stakeholders

In occupational health care, it is essential to involve a wide variety of stakeholders to the work. Various actors are to be taken into consideration, as occupational health care addresses many important issues. The continuous interaction with the various stakeholders is essential, and the active work with the stakeholders can be seen important in reducing uncertainty of the operating environment by identifying and following important actors and critical dependencies [27]. Furthermore, understanding the mutual dependencies and 'co-destinies' of the various actors is critical in an modern-operating environment. An essential factor to be taken into consideration is the potential impact of the multiple and conflicting stakeholder interests [33].

Comparative studies have shown that one success factor of occupational health interventions in organizations is the establishment and working of steering groups involving key stakeholders [34].

The effective stakeholder work requires also the recognition of the different roles of different stakeholders. Stakeholders differ according to their resources. In organizational contexts, where stakeholders are active, knowledgeable and inter-dependent, success is dependent on active, practical stakeholder relationship management [35, 36].

The diversity of the stakeholders has led to various classification structures. In this article, the starting point is the model developed by Fottler et al. for the stakeholders of hospitals. Their classification included three layers: internal stakeholders, interface stakeholders and external stakeholders [19].

It is generally understood, that in healthcare organizations the primary stakeholders are the patients [37]. Innovative pharmaceutical companies, such as Novo Nordisk in their diabetes drug development, have also placed the user in the middle of their stakeholder chart [10] (see **Figure 1**).

Another key internal stakeholder group is the management of an organization. The management includes senior management, line management and human resources management (HRM). All these groups and their collaboration are essential for successful occupational health care [34, 38, 39].

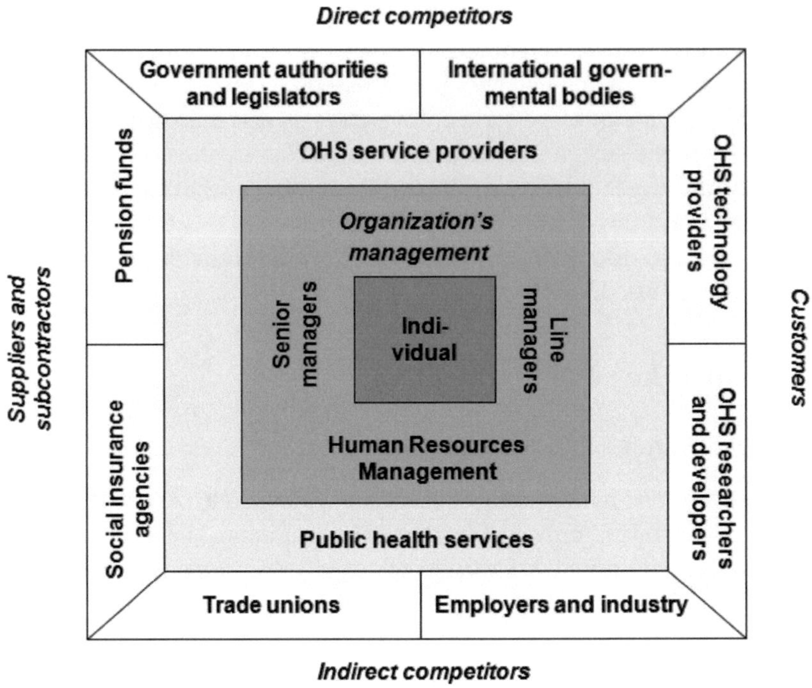

Figure 1. Key stakeholders of occupational healthcare (modified with permission from [14]).

The interface stakeholders are the occupational health-service providers, who provide their services to organizations and also take part in the development of occupational health care [40]. In addition, the public healthcare providers can also be seen as interface stakeholders (see e.g. [41]).

The external stakeholders consist of all the other important stakeholders, such as government and governmental agencies, trade unions, employers' unions, pension funds, municipalities, technology providers and service providers. Furthermore, other external stakeholders also, such as customers and suppliers of an organization as well as shareholders and competitors can be regarded as external stakeholders of occupational health care.

3.2. Internal stakeholders

The key rationale of occupational health care is to provide a safe working environment for the individuals and also to promote their health and well-being (see e.g. [6, 12, 34]). In this article the key stakeholder is an individual (employee) working in an organization, and thus they are in the centre of the stakeholder chart.

Various studies on occupational health care show that strong participation of the employees and utilization of participatory principles in the occupational health care interventions are critical success factors (see e.g. [34]). The wide variation in the health conditions of the individuals poses challenges to occupational health care activities in the workplace. For instance, disparities in living habits (including nutritional habits, exercising habits etc.) are wide within working age population in many countries and also within workplaces. It is also notable that these disparities are also partially linked to the socio-economic status of the individual, and thus are also out of the reach of the mandate of occupational health care.

Within the work environment, the occupational health care actions can also be supported by peer groups – in particular, in workplace health promotion (see e.g. [42]) – which can also be considered as important internal stakeholders.

The managers of an organization play an important role in occupational health care. Their decisions of policy guidelines, resource allocations as well as daily management define the destiny of occupational health care activities. Thus various managers must be regarded as major internal stakeholders (see e.g. [39]). For instance, a Canadian study showed that general managers and human resource managers differed in their attitudes and ambitions regarding workplace health promotion [43]. It is important to regard several managers, who have different roles in an organization, of different management levels as important internal stakeholders of occupational health care.

Senior management support for occupational health care has been seen in many studies as a critical success factor (see e.g. [34]). Recent comparative study showed, that there is considerable evidence of the negative impact of senior management support to occupational health care interventions, but less evidence about the positive impact of the management support component of intervention processes [34].

The middle managers are the actual key drivers of occupational health care work and thus their role is imperative within an organization (see e.g. [34, 38, 44]). The middle managers are the drivers of change also in this area, but they can also block important processes within a workplace [34].

Human Resources Management (HRM) is supporting both senior management and middle managers, and their support for the occupational health care activities for managers in formulating goals and implementing action plans are considerable [38, 45].

3.3. Intermediate stakeholders

The essential intermediate stakeholders for occupational health care can be defined to be occupational health services (OHS) providers and generic public healthcare providers.

The contemporary requirements for OHS providers are multifaceted ranging from health promotion of individuals to the development of the working environment (see e.g. [40]). It is also obvious that many employers do not have sufficient knowledge of occupational health care [40], and thus the OHS service providers can be strategically also an important partner in developing the working environment of an organization. The benefits of such a strategic partnership also require a lengthy and open process of collaboration between an employer and an OHS provider.

It is also critical to regard public health care service providers as important stakeholders of occupational health care, as there exist important differences in scope and options for various interventions between public health and workplace health contexts [46]. Although a country might have a well-working public health system, the organizations are willing to develop and widen their occupational health care activities (see e.g. [43]). Understanding the potential and services of public health care can also focus on the occupational health services more effectively. Unfortunately, according to some studies, the cooperation between general practitioners and occupational health physicians is often lacking or sub-optimal [41].

3.4. External stakeholders

The legislative framework of the occupational health care has been developing during the last decades in most industrialized countries. The legal requirements create the necessary framework for occupational health care, and they are elementary in encouraging the employer engagement to the provision and development of occupational health care [47].

The development of the legislative framework, laws and regulations can also be initiated by international global organizations (such as the WHO or ILO, see e.g. [6]). For member states of the European Union, the policy and regulation development within the EU can be an important accelerator for occupational health care (see e.g. [48]).

Trade unions are important stakeholders, as they represent the interests of employees in an organization (see e.g. [12]). Furthermore, the interests of trade unions with regard to occupational health care can have a different emphasis from those of employers. Trade unions have an important role in improving workplace conditions. In a recent Finnish study on stakeholder positions of occupational health care, trade unions seem to fear that health promotion programs distract attention from workplace health hazards [14].

The organizations and conglomerates of employers have clearly understood the challenges in occupational health care, and they also understand the value of joint development actions to tackle common problems, such as increasing absenteeism and rising trend of problems linked to mental health challenges for employees (see e.g. [14]).

Pension funds are also important stakeholders in occupational health care. They do have an incentive to prevent disability and early retirement, since they have to pay a large part of the eventual pension expenses [49].

Social insurance agencies and other public authorities can benefit from well-organized and systematic cooperation between organizations, OHS providers and social insurance offices. A Swedish study showed the clear financial and operational benefits of systematic and solution-oriented co-operation for all parties involved [50].

Developers of new technological solutions for occupational health care can be interesting stakeholders, as they can provide novel solutions e.g. for workplace health promotion (see e.g. [22]). However, the challenges in such new technologies and services can often be, that the buyers, users and payers might be separate entities with disparate interests [51].

The role of researchers and developers of occupational health care can also be important for organizations. In some countries, like Finland, there are specialized research institutes for occupational health care [14]. The close collaboration between research institutions, universities and companies can be essential in developing novel solutions and methods for occupational health care (see e.g. [52]).

A wide range of various external stakeholders are pushing organizations to improve their work and respond in more responsible ways to contemporary challenges (see e.g. [53]). Thus important occupational health care stakeholders are also the suppliers, customers and competitors of an organization. Well-organized occupational health care can be an important element of the image of an organization, as it competes of its customers as well as of highly-skilled professional workforce [12, 23].

4. Analysing key positions of stakeholders

4.1. Researching stakeholders

After the thorough identification on key stakeholders, it is essential to understand their key positions, arguments and interests in occupational health care work.

It is necessary to recall that the function of time is also an important issue to consider in stakeholder analysis, as stakeholder interests may converge over time. Stakeholder positions are not, however, static. Rather it is clear that stakeholder positions change and evolve over time [10].

A variety of methods can be used to map the opinions and positions of various stakeholders. Recommendable methods include face-to-face interviews using checklists, semi-structured interviews and structured questionnaires, which all can be used to collect data from primary sources [26]. Secondary methods could include analysis of published and unpublished documents, policy statements and various regulations [26]. An effective method can also be the utilization of focus groups, which might also be venues for development of novel ideas and tackling complex issues (see e.g. [54]). During the planning phase of stakeholder analysis, the strengths and weaknesses as well as the resource requirements of various methods should be clarified [54].

During a Finnish case study of stakeholder positions in workplace health promotion, 45–60-min long semi-structured, thematic interviews undertaken by two researchers was the main method used. The interviews were chosen as a key information collection method, as they also enabled the capturing of detailed knowledge, potential clarifications and the amplification of earlier questions [14].

4.2. Mapping and assessing stakeholder positions

The wide literature of stakeholder analysis provides many different examples of matrices, charts, position maps, network maps and other figures to present the data analysed and collected [26]. For the purposes of this article, two tools are used to illustrate the stakeholder positions. These are the division to primary and secondary stakeholders, and the power/influence matrix. These tools should be seen as complementary to each other.

According to the well-known definition of Clarkson, the "primary stakeholder group is one without whose continuing participation the corporation cannot survive as going concern" [31]. Furthermore, the "secondary stakeholder groups are defined as those who influence or affect, or are influenced or affected by, the corporation, but they are not engaged in transactions with the corporation and are not essential for its survival" [31].

In the context of occupational health care, the classification into primary and secondary stakeholders shows the fundamental understanding of an organization of the key target groups of its occupational health care work, and which stakeholders it needs to take into consideration in fulfilling the objectives of its occupational health care work. Thus the primary stakeholders are the elementary groups to work with and the secondary stakeholders are necessary and required supporting groups.

A simplified visualization is to present the primary and secondary stakeholders of occupational health care in a two-column table which also includes short commentaries of their importance (see **Table 1**).

Primary stakeholders	Secondary stakeholders
Stakeholder A	Stakeholder G
- rationale 1	- rationale 11
- rationale 2	- rationale 12
Stakeholder B	Stakeholder H
- rationale 3	- rationale 13
- rationale 4	- rationale 14
Stakeholder C	Stakeholder I
- rationale 5	- rationale 15
- …	- …

Table 1. Classification to primary and secondary stakeholders.

However, the classification to primary and secondary stakeholders does not yet indicate the relative power, influence or interest of the stakeholders – this can be achieved by undertaking an evaluation of the power and interest of the various stakeholders. This is usually visualized with a power/interest matrix showing the assessment of the organization of the contemporary importance of the various stakeholders regarding occupational health care.

4.3. Power and interest of stakeholders

The implementation and development of effective occupational health care within an organization requires support from several key stakeholders. Thus it is also important to assess their relative power and influence. This is undertaken by reporting the stakeholder positions or power and influence, which can also be visualized as a power/interest matrix.

In the power/interest matrix there are two important sets of questions to be assessed. According to the classification proposed by Johnson and Scholes, the question "If we were to pursue this strategy with disregard to the views of this particular stakeholder, could/would they stop it?" assesses the power of the stakeholder. The interests of the stakeholder is assessed with the questions "How high is this approach on their priorities?" and "Are they likely to actively support or oppose this approach, or will their interest be short-lived?" [32].

The power dimension indicates the level of influence a stakeholder has in either supporting or resisting a strategic initiative. Stakeholders may exercise their power in many ways, for example through a legal position, possession of knowledge and key resources or even informal networking with other decision makers. The interest dimension depends on how high a priority this strategy is. Interests can be open or hidden, which makes their assessment challenging. Interests may be based on a stakeholder's anticipated economic gain, brand value or power position. The level of interest can be estimated by assessing whether a stakeholder has a long-term commitment to the strategy [32].

In a recent Finnish case study on stakeholder views on workplace health promotion, the researchers asked the 17 interviewees to describe their position among the stakeholders and also to assess the other stakeholders and their relative power and interest (see **Table 2**). The interviewees were asked to present arguments about the power and of the interest of the other

Stakeholder	Power/influence	P score	Interest	I score
Individuals (active)	Responsibility for the individual's own health is a prerequisite for the success of proactive health measures. Active individuals are promoters of healthy lifestyles. Active individuals will also use personal financial resources to promote their individual health.	Medium	There is a large disparity between groups of different socioeconomic status. The number of active individuals is growing. They can see the benefit in improved personal and working ability.	High
Individuals (passive)	No contribution to personal or corporate wellness development activities.	Low	Lacking interest, rather resistant towards healthy lifestyles. Need external motivation to participate in health programs.	Low
Occupational Health Service (OHS) providers	OHS providers are seeking more strategic partnerships with their customers. Potential linkages between pension funds and OHS providers could eventually lead to a new landscape for preventive OHS.	Medium	Current business models are based on charges per services–an alternative 'service contract' model could provide incentive for proactive health measures.	High
Pension funds	Pension funds have huge financial resources at hand, e.g. to prevent early retirement by improving the health conditions of the insured persons. Pension funds have power over vocational rehabilitation decisions.	Medium	Considerable benefits in avoiding expenses gained from the prevention of disability and early retirement, Value-added services are critical in the competition between pension funds, because in Finland they cannot compete with prices and other financial benefits.	High
Technology providers	Enablers of new technology-based business models for occupational health care.	Low	Occupational health services could potentially become a new market for technology products.	Medium

Table 2. Example of a Finnish study of stakeholder positions of workplace health promotion – a modified excerpt (applied from [14]).

stakeholders and these were summarized. Based on the interviews, the researchers assessed each stakeholder identified in the stakeholder typology according to their power and interest using a three-grade score (high, medium and low) [14]. This score was also used in building a power/interest matrix.

The results of the power/interest matrix can also be summarized as a two-dimensional power-interest matrix (see **Table 3**). Stakeholders in the upper right corner (high/medium power and interest) are the key players in driving the change towards effective occupational health care (see also e.g. [14]).

POWER			
High			
Medium			
Low			
	Low	Medium	High
		INTEREST	

Table 3. Power / interest matrix.

5. Developing stakeholder strategies

The comprehensive stakeholder analysis is an important tool in identifying, planning and implementing strategies for managing stakeholder relationships and identifying current and future opportunities and threats (see e.g. [26]).

The results of the work of stakeholder analysis presented in the power/interest matrix are valid for establishing guidelines with regards for the work with various stakeholders (see **Table 4**). Naturally it should be understood that such a classification can only give rough guidance, as in every case the stakeholders vary from another in their activity, involvement and energy. However, such a simplified classification can ensure that all the stakeholders are regarded and that no identified stakeholder groups are left unnoticed.

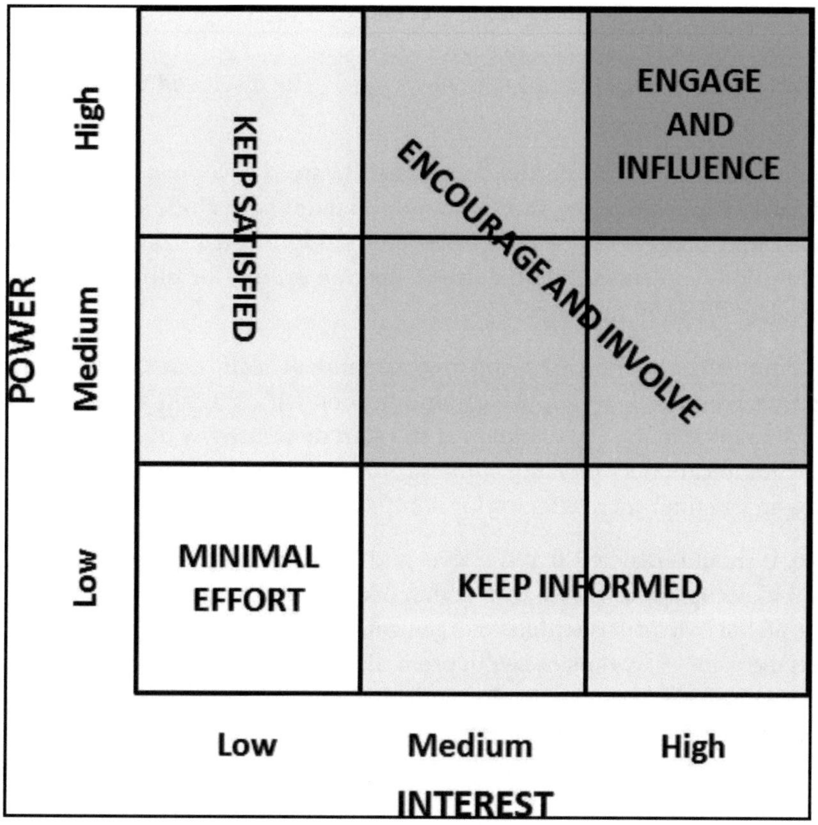

Table 4. Power / interest matrix and strategies according to stakeholders.

It is obvious that not much effort should be directed on stakeholders with little power and a low level of interest. It is good to note their presence, but no particular actions are needed for them.

The needs of stakeholders with low power, but high or medium interest should be addressed mainly through continuous, selected information distribution. Gaining the support of these stakeholders through lobbying can be a good tactic, because they can be valuable allies in influencing the attitudes of other, more powerful stakeholders (see e.g. [55]).

Stakeholders with high or medium power, but low interest, are often difficult to plan with and to develop consistent strategies. These stakeholders might, in general, be quite passive, but might unexpectedly exercise their power in reaction to a particular event or policy. Under-estimation of this group can have disastrous consequences for the adoption of the new approach. These stakeholders should be kept satisfied through continuous communication, and possibly also through selected involvement to focal activities.

Stakeholders in the middle of the power/interest matrix with relatively high power and interest should be encouraged for a solid and continuous support of the work and activities undertaken. These stakeholders can be valuable resources and also provide required support to plan and initiate new ideas.

The most important stakeholders, who are crucial to the success of any strategic development in occupational health care, are the ones with high power and with high interest. These might also be stakeholders, whose opinions and views need to be discussed and elaborated, as their views can also differ and vary.

One recommended method in occupational health care development is to utilize steering groups, which should include key stakeholders, and thus also include key agents for the work to be planned and undertaken. A comprehensive study of organization-level occupational health interventions concluded that the use of steering groups for projects was unanimously recommended by different developers [34].

Another recommended method is having for occupational health care clearly an owner within an organization, who also keeps up the communication with selected stakeholders. It is counterproductive to involve the stakeholders at the start of an activity and after that leave them out of the communication loop. Many studies confirm that active communication with different parties is an essential success factor for occupational health care activities (see e.g. [34]).

Furthermore, it should be noted that the level and intensity for participation to the work and development of occupational health care within one stakeholder group may vary significantly. For instance, all the individuals within an organization might not share the same ambitions and do not assess the value of workplace health promotion in a similar manner. Thus when developing strategies for stakeholder participation, it is also important to understand that the range of participation can be wide from passive participation to self-directed activities and work.

6. Summary

Occupational health care today consists of the combined efforts of employers, employees and other stakeholders to improve the health and well-being of people at work. This conceptual development has meant that the number of stakeholders in occupational health care is increasing. Modern successful occupational health care requires multifaceted collaboration, partnerships and alliances between various actors.

An essential element in occupational health care is well-structured work with various stakeholders. The work with stakeholders starts with stakeholder analysis. It aims to evaluate and understand stakeholders from the perspective of an organization. The first step is to identify the essential stakeholders of occupational health care in an organization. The stakeholders can be classified as internal, intermediate and external stakeholders. The most important internal stakeholders are the individual employees working in an organization; other internal stakeholders are peer employees and various managers (senior managers, line managers, HR managers) of an organization. Interface stakeholders include occupational health service

http://dx.doi.org/10.5772/66479

providers and public healthcare providers. A wide range of various external stakeholders are pushing organizations to improve their work and respond in more responsible ways to contemporary challenges.

After identifying key stakeholders, the next step is to understand their key positions, arguments and interests in occupational health care work. A variety of methods can be used to map the opinions and positions of various stakeholders. It is important to summarize the views of the stakeholders and also analyse their relative positions according to power, influence and interest.

The stakeholder analysis should lead to planning and implementing strategies for managing stakeholder relationships and utilizing the full potential of various stakeholders in developing occupational health care within an organization.

Author details

Ari-Matti Auvinen

Address all correspondence to: ama.auvinen@aalto.fi

HEMA Institute, Aalto University, Espoo, Finland

Human Capital Investment Oy, Helsinki, Finland

References

[1] Husman K, Husman P. Human resource needs of modern occupational health services—the case in Finland. Scand J Work Environ Health Suppl. 2005;1:46–50.

[2] Chu C, Breucker G, Harris N, Stitzel A, Gan X, Gu X,et al. Health-promoting workplaces—international settings development. Health Promot Int. 2000;15(2):155–67.

[3] Peltomäki P, Johansson M, Ahrens W, Sala M, Wesseling C, Brenes F, et al. Social context for workplace health promotion: feasibility considerations in Costa Rica, Finland, Germany, Spain and Sweden. Health Promot Int. 2003;18(2):115–26.

[4] Peltomaki P, Husman K. Networking between occupational health services, client enterprises and other experts: difficulties, supporting factors and benefits. Int J Occup Med Environ Health. 2002;15(2):139–46.

[5] Stoney C, Winstanley D. Stakeholding: confusion or utopia? Mapping the conceptual terrain. J Manag Stud. 2001;38(5):603–26.

[6] Burton J, World Health Organization. WHO Healthy workplace framework and model: background and supporting literature and practices. Geneva: World Health Organization; 2010. 123 p.

[7] Freeman RE. Strategic management: a stakeholder approach. Boston: Pitman Publishing; 1984. 275 p.

[8] Preston LE, Sapienza HJ. Stakeholder management and corporate performance. J Behav Econ. 1990;19(4):361–75.

[9] Bryson JM. What to do when stakeholders matter. Public Manag Rev. 2004;6(1):21–53.

[10] Freeman, RE, Harrison, JS, Wicks, AC. Managing for stakeholders. New Haven: Yale University Press; 2007. 200 p.

[11] Maignan I, Ferrell OC, Ferrell L. A stakeholder model for implementing social responsibility in marketing. Eur J Market. 2005;39(9/10):956–77.

[12] Amponsah-Tawiah K, Mensah J. Harmonising stakeholder interests: the role of occupational health and safety. Afr J Business Manag. 2015;9(9):394–401.

[13] Donaldson T, Preston LE. The stakeholder theory of the corporation: concepts, evidence, and implications. Acad Manag Rev. 1995;20(1):65–91.

[14] Auvinen AM, Kohtamäki K, Ilvesmäki, A. Workplace health promotion and stakeholder positions: a Finnish case study. Arch Environ Occup Health. 2012;67(3):177–84.

[15] Zinkhan GM, Balazs AL. A stakeholder-integrated approach to health care management. J Business Res. 2004;57(9):984–9.

[16] Murdock A. Stakeholder theory, partnerships and alliances in the health care sector of the UK and Scotland. Int Public Manag Rev. 2004;5(1):21–40.

[17] Keele RL, Buckner K, Bushnell S. Identifying health care stakeholders: a key to strategic implementation. Health Care Strateg Manag. 1987;5(9):4–10.

[18] Daake D, Anthony WP. Understanding stakeholder power and influence gaps in a health care organization: an empirical study. Health Care Manag Rev. 2000;25(3):94–107.

[19] Fottler MD, Blair JD, Savage GT, Whitehead CJ, Laus MD. Assessing key stakeholders: who matters to hospitals and why?. Hosp Health Serv Admin. 1989;34(4):525–47.

[20] Wells R, Lee SY, McClure J, Baronner L, Davis L. Strategy development in small hospitals: stakeholder management in constrained circumstances. Health Care Manag Rev. 2004;29(3):218–28.

[21] Tjulin Å, Stiwne EE, Ekberg K. Experience of the implementation of a multi-stakeholder return-to-work programme. J Occup Rehab. 2009;19(4):409–18.

[22] Reijonsaari K. Co-creating health-examining the effects of co-creation in a lifestyle intervention service targeting physical activity. Espoo: Aalto University; 2013. 166 p.

[23] Ramirez R. Value co-production: intellectual origins and implications for practice and research. Strateg Manag J. 1999;20(1):49–65.

[24] Auvinen AM. Personal health systems and value creation mechanisms in occupational health care. In: Proceedings of the 29th Annual International Conference of the IEEE EMBS; 23-26 Aug 2007; Lyon. New York: IEEE. p. 5882–5885.

[25] Brugha R, Varvasovszky Z. Stakeholder analysis: a review. Health Policy Plan. 2000;15(3):239–46.

[26] Varvasovszky Z, Brugha R. A stakeholder analysis. Health Policy Plan. 2000;15(3):338–45.

[27] Yläranta M. Between two worlds: stakeholder management in a knowledge intensive governmental organisation. Publications of the Turku School of Economics, Series A-7:2006. Turku: Turku School of Economics; 2006. 188 p.

[28] Simmons J, Lovegrove I. Bridging the conceptual divide: lessons from stakeholder analysis. J Organ Change Manag. 2005;18(5):495–513.

[29] Freeman RE, Reed DL. Stockholders and stakeholders: a new perspective on corporate governance. California Manag Rev. 1983;25(3):88–106.

[30] Mitchell RK, Agle BR, Wood DJ. Toward a theory of stakeholder identification and salience: defining the principle of who and what really counts. Acad Manag Rev. 1997;22(4):853–86.

[31] Clarkson ME. A stakeholder framework for analyzing and evaluating corporate social performance. Acad Manag Rev. 1995;20(1):92–117.

[32] Johnson G, Scholes K. Exploring corporate strategy—text and cases. 5th ed. Great Britain: Prentice Hall Europe; 1999. 588 p.

[33] Smith AM, Fischbacher M. New service development: a stakeholder perspective. Eur J Market. 2005;39(9/10):1025–48.

[34] Nielsen K, Randall R, Holten AL, González ER. Conducting organizational-level occupational health interventions: What works?. Work Stress. 2010;24(3):234–59.

[35] Peltokorpi A, Alho A, Kujala J, Aitamurto J, Parvinen P. Stakeholder approach for evaluating organizational change projects. Int J Health Care Qual Assur. 2008;21(5):418–34.

[36] Savage GT, Nix TW, Whitehead CJ, Blair JD. Strategies for assessing and managing organizational stakeholders. Acad Manag Exec. 1991;5(2):61–75.

[37] Werhane PH. Business ethics, stakeholder theory, and the ethics of healthcare organizations. Camb Q Healthc Ethics. 2000;9(02):169–81.

[38] Larsson R, Stier J, Åkerlind I, Sandmark H. Implementing health-promoting leadership in municipal organizations: managers' experiences with a leadership program. Nordic J Work Life Stud. 2015;5(1):93.

[39] Tappura S, Syvänen S, Saarela KL. Challenges and needs for support in managing occupational health and safety from managers' viewpoints. Nordic J Work Life Stud. 2014;4(3):31.

[40] Schmidt L, Sjöström J, Antonsson AB. Successful collaboration between occupational health service providers and client companies: key factors. Work. 2015;51(2):229–37.

[41] Moßhammer D, Natanzon I, Manske I, Grutschkowski P, Rieger MA. Cooperation between general practitioners and occupational health physicians in Germany: how can it be optimised? A qualitative study. Int Arch Occup Environ Health. 2014;87(2):137–46.

[42] Shain M, Kramer DM. Health promotion in the workplace: framing the concept; reviewing the evidence. Occup Environ Med. 2004;61(7):643–8.

[43] Downey AM, Sharp DJ. Why do managers allocate resources to workplace health promotion programmes in countries with national health coverage?. Health Promot Int. 2007;22(2):102–11.

[44] Mellor N, Webster J. Enablers and challenges in implementing a comprehensive workplace health and well-being approach. Int J Workplace Health Manag. 2013;6(2):129–42.

[45] Hasson H, Villaume K, von Thiele Schwarz U, Palm K. Managing implementation: roles of line managers, senior managers, and human resource professionals in an occupational health intervention. J Occup Environ Med. 2014;56(1):58–65.

[46] Karanika-Murray M, Weyman AK. Optimising workplace interventions for health and well-being: a commentary on the limitations of the public health perspective within the workplace health arena. Int J Workplace Health Manag. 2013;6(2):104–17.

[47] Martinsson C, Lohela-Karlsson M, Kwak L, Bergström G, Hellman T. What incentives influence employers to engage in workplace health interventions?. BMC Public Health. 2016;16(1):854.

[48] Leka S, Jain A, Zwetsloot G, Cox T. Policy-level interventions and work-related psychosocial risk management in the European Union. Work Stress. 2010;24(3):298–307.

[49] Piekkola H. Active ageing policies in Finland. Keskusteluaiheita; Discussion papers; No. 898. Helsinki: Etla; 2004. 42 p.

[50] Kärrholm J, Ekholm K, Ekholm J, Bergroth A, Ekholm KS. Systematic co-operation between employer, occupational health service and social insurance office: a 6-year follow-up of vocational rehabilitation for people on sick-leave, including economic benefits. J Rehabil Med. 2008;40(8):628–36.

[51] Ilvesmaki A. Drivers and challenges of personal health systems in workplace health promotion. In: Proceedings of the 29th Annual International Conference of the IEEE EMBS; 23–26 Aug 2007; Lyon. New York: IEEE. p. 5878–5881.

[52] Stone A, Usher D, Marklin R, Seeley P, Yager JW. Case study for underground workers at an electric utility: How a research institution, university, and industry collaboration improved occupational health through ergonomics. J Occup Environ Hyg. 2006;3(8):397–407.

[53] Waddock SA, Bodwell C, Graves SB. Responsibility: the new business imperative. Acad Manag Exec. 2002;16(2):132–48.

[54] Reed MS, Graves A, Dandy N, Posthumus H, Hubacek K, Morris J, Prell C, Quinn CH, Stringer LC. Who's in and why? A typology of stakeholder analysis methods for natural resource management. J Environ Manag. 2009;90(5):1933–49.

[55] Auvinen AM, Waddington L, Moretti M, Dondi C, Fischer T, Kretschmer T, et al. Understanding the stakeholders—key to successful implementation of adult learning projects. eLearning Papers Special edition 2010. 2010:1–15.

Chapter 3

Occupational Risks of Health Professionals

Nilgun Ulutasdemir and Ferdi Tanir

Additional information is available at the end of the chapter

http://dx.doi.org/10.5772/67148

Abstract

Health service is an important work area which can lead to important risks related to occupational health and safety (OHS) of employees. This book chapter is prepared to evaluate the effects of occupational risks on health and decrease the exposure to occupational risks of health professionals by searching national and international literatüre. Thus, awareness can be raised to define occupational risks and help planning services for health professionals. American National Institute for Occupational Safety and Health (NIOSH) has reported 29 kinds of physical, 25 kinds of chemical, biological 24 varieties, 10 and six kinds of ergonomic and psycho-social hazards and risks. According to ILO, it has been reported that there is 1.25 trillion dollars loss each year due to the OHS problems. In Turkey, the loss of only social security systems has been reported as approximately 4 million Turkish Liras per year. Health professionals have work stress, and they suffer from the inconvenient design and the hazards within the workplace. The health of the health professionals affects the health of the community. Thus, it is important to decrease the exposure to occupational risks of health professionals and diligently work on this issue.

Keywords: health, safety, occupational risk factors, occupational management, health workers

1. Introduction

Health service is an important work area which can lead to important risks related to health and safety of employees [1]. Occupational risks of health professionals can alter according to the profession, the work itself and the unit of the hospital [2].

Work conditions of health professionals are quite challenging in some regions. The precautions about the occupational safety are not sufficient in hospitals and health institutions. Professional development and education opportunities of health professionals are limited, and the professional organizations are also insufficient [3]. It has been reported that increasing working hours per week results in an increase in the possibility of being injured. Professionals, who did not receive the occupational health and safety (OHS) training before, have more accidents [4]. Nonfatal accidents at work and occupational disease cases of health professionals take place at the top compared to other industries [5]. But in Turkey, the inclusion of hospitals in "Very Dangerous Jobs" class was approved barely in 2009 via "Hazard Classes List Notification Relating Work Health and Safety" [6, 7]. This list was formed according to NACE code.

2. Occupational risks of health professionals

The health-related risks associated with health professionals can be grouped as psychosocial, physical, biological, chemical and ergonomic factors. American National Institute for Occupational Safety and Health (NIOSH) has reported 29 kinds of physical, 25 kinds of chemical, biological 24 varieties, 10 and six kinds of ergonomic and psycho-social hazards and risks [8–11].

3. Physical risk factors

The main physical risk factors which affect health professionals are ionizing and nonionizing radiation, noise, lighting, electrical assembly, slippery floors, hot/cold, ventilation, vibration and indoor pollution. Ionizing radiation is one of the most important physical hazards in hospitals, and it influences various different health professionals in different units and with different professions (mainly radiotherapy, nuclear medicine and radiology staff). It has carcinogenic, teratogenic and mutagenic impacts, and it is fatal in high concentrations. It leads to burns, cataracts, infertility, genetic and congenital anomalies in moderate concentrations, and it causes cancer (particularly leukemia) during long-term exposure [8, 12, 13]. Nonionizing radiation is another physical risk factor for health professionals. It has been stated that the increment in the use of devices with electromagnetic fields and exposure to these electromagnetic fields deteriorate the body balance and lead to diseases. It has been reported that the exposure to nonionizing radiation particularly during the work time about 8–10 h leads to feeling of dryness in the throat, eye problems, headaches, allergies, facial flushing, insomnia, sensitivity to sounds, hearing difficulties and fatigue [8, 14, 15].

Loudness is another important factor, which disturbing people, complicates the communication, restricts the relaxation, adversely affects and harms the nervous system, reduces the work efficiency and creates hearing problems. Studies have shown that loudness has increased at a level of discomfort in patients and health professionals in hospitals in the last 50 years [12, 16, 17].

The ventilation system of the hospitals has importance in the health protection of both patients and health professionals in terms of nosocomial infections. Therefore, the ventilation systems should be established to prevent the nosocomial infections by paying attention to biological and physical features of related microorganisms [2].

Another physical factor that can affect the health professionals is the lighting of the workplace. A sufficient and satisfactory level of lighting should be arranged in order to provide a comfortable workplace to the health professionals [18].

4. Chemical risk factors

Various chemicals are key agents which are used in order to diagnose and treat the diseases, perform the preventive applications and take hygienic precautions, whereas they are hazardous for health status of health professionals. Health professionals are exposed to chemicals (disinfectants to anesthetic agents, cytotoxic agents, drugs and some heavy metals such as mercury and latex) repeatedly and sometimes in very dense amounts. Similarly, their impacts show a wide variety according to the concentrations, contact time and way, the presence of other risky agents and personal features, etc. [19]. Acids and alkalis, salts, dyes, volatile organic solvents, various drugs including primarily anticancer drugs in pathology, biochemistry, hematology and other laboratories are important risk factors for a series of diseases from allergy to cancer [8, 20].

Drugs that cause severe organ toxicity and other toxic effects and drugs that show mutagenic, carcinogenic, teratogenic effects or any of the reproductive system disorders are defined as "hazardous drugs." The long-term exposure to these antineoplastic/cytotoxic drugs used in chemotherapy leads to potential risks in health professionals. At the stage of preparation, administration and waste disposal of these drugs, severe health outcomes can be observed due to the inhalation of powder and droplets, the absorption through the skin, the ingestion of contaminated food as well as particularly teratogenic, carcinogenic and genotoxic effects that threaten the reproduction during pregnancy [10, 12, 13, 19–23].

5. Biological risk factors

5.1. Agents transmitted by respiration and droplets

Some agents such as droplets and droplet cores can be transmitted via respiratory secretions of patients. Tuberculosis, measles, rubella, chicken pox, *severe acute respiratory syndrome* (SARS), influenza, meningococcal and pneumococcal infections transmit in this way [10, 13]. According to the various studies performed in Turkey, it has been detected that health professionals, particularly nurses, are under the risk particularly those working in pulmonary disease services [24, 25]. Demir et al. have performed a study in order to determine the tuberculosis infection risk among health professionals working in pulmonary disease hospital and another hospital which does not have pulmonary disease clinic. They have shown that the tuberculosis infection risk was 7.4 times higher in pulmonary disease hospital compared to the other hospital without pulmonary disease clinic because of the higher tuberculosis exposure [26].

5.2. Infections transmitted through direct contact

These infectious agents transmit through direct contact with the patient. There is no need to be in contact with the skin or mucosa as well as the loss of skin integrity for the transmission.

Resistant bacteria and skin parasites such as scabies are examples of microorganisms which can lead to severe infections in inpatients [24].

5.3. Biological agents transmitted by blood and bloody body fluids

These biological agents can transmit through the skin due to its impaired integrity and mucous membranes (mouth, eye and urogenital mucosa) as a result of the exposure to blood and/or bloody body fluids and some sterile body fluids. Even though there are almost 30 microorganisms which can be transmitted in this way, the most important ones are hepatitis B virus (HBV), hepatitis C virus (HCV), hepatitis D virus (HDV) and human immunodeficiency virus (HIV) since they can lead to systemic infections because of their current importance. The diversity of the clinical outcomes of these agents varies from asymptomatic infections to severe and even fatal infections [3, 4, 12, 13, 24, 27, 28]. The transmission of the infections via blood occurs mostly by the penetration of the needles used in patients, injury with contaminated sharp instruments or mucosal splashes infected blood or body fluids [29, 30]. It has been specified that the two-third of the health professionals stated that they were exposed to blood and/or body fluids at least once, the HIV infection was related to the profession in the 57% of the HIV-positive health professionals, and the risk of developing hepatitis B infection in health professionals is 10 times more compared to the general population [12, 13]. For example, in Turkey, you can see the Crimean-Congo Hemorrhagic Fever (CCHF) and Margburg which make hemorrhagic fever and which are infected with blood.

6. Ergonomic factors

The more the harmonization is ensured between employer and work environment, the better the safety and efficiency can be provided to employees [31]. In case there is an inconsistency between the physical capacity of the laborer and the physical requirements, occupational diseases may occur [32]. Particularly, nurses are in the third row after heavy industry workers and the heavy vehicle drivers who can experience the musculoskeletal system problems [8, 33]. According to the study performed in Turkey in which the prevalence of low back pain in the last 12 months and related factors were examined, it has been shown that the prevalence of low back pain in the last 12 months was at a high level that affects the working life (73.3%) [34]. In another study, we have indicated that the frequency of the low back pain in the past 1 year was 58.3% in nurses and health officers and 33.0% in the sick nurses [35].

7. Psychosocial factors

Job satisfaction explains the harmony between the expectations of the employee from the institution and the profession and the opportunities offered to employee. The job satisfaction is one of the factors, which affects the burnout which has severe outcomes particularly in professions which provide the service directly to human. An intense burnout state can lead to problems such as resigning, incompatibility in marriage and family, decreased in self-esteem, a difficulty to concentrate and social isolation [12]. One of the factors that adversely influence the job satisfaction and lead to burnout in the workplace is violence. A violence, which has affected the health of health professionals recently, can be observed as a problem in every health institution and health professional [8]. The violence in a health institution is defined as a verbal or behavioral threat, physical or sexual assault [36]. It has been increasingly more accepted that the health professionals, who must contact directly with individuals in difficult situations due to their deteriorated health status, are the most important target and victim of the occupational violence among all professions [37, 38]. For instance, in Turkey, totally five doctors were killed in the past 10 years due to the violence in hospitals. The names of the doctors were as follows: Prof. Dr. Göksel Kalayci (11th of November, 2005), Dr. Ali Menekse (15th of January, 2008), Dr. Ersin Arslan (17th of April, 2012), Dr. Melike Erdem (30th of November, 2012), Dr. Kamil Furtun (29th of May, 2015) and Aynur Dagdemir (19th of November, 2015) [2]. It has been estimated in some studies that the risk of exposure to violence of particularly health professionals by patients, families of patients or others is 4–16 times more compared to other employees in various sectors (such as guards, police officers, bank employees, retail workers, those working in the transport sector) [39].

According to the 2002 report of the World Health Organization (WHO) entitled "Workplace violence in the health sector," the International Labor Organization (ILO) and the International Union of Nurses (ICN), it has been reported that more than 50% of the health professionals are exposed to violence [40]. In this report, generally 3–17% of the health professionals were exposed to physical, 27–67% of them were exposed to verbal, 10–23% of them were exposed to psychological, 0.78% of them were exposed to sexual and 0.8–2.7% of them were exposed to ethnical violence [41]. The violence has long-term impacts on health professionals such as despondency, job loss, discontentedness, decreased job satisfaction, anxiety, life-threatening injury, restlessness, anger, stress disorder, nightmares and sleep problems. The violence in the workplace not only affects the employee but also influences the colleagues, family and friends of the individual [42–47].

The studies performed on the workplace violence have shown that the psychological violence has reached to more dangerous levels compared to physical violence. Psychological violence in the workplace (mobbing) is defined as systematic and hostile attitudes of one or several people toward one individual with unethical communication. Mobbing is performed generally by management team, but it can be directed also by colleagues, subordinates or a group of employees. It has been specified that health professionals particularly nurses are seriously a risk group exposed to mobbing in workplace [48, 49]. Mobbing leads to excessive stress, exclusion, anxiety, digestive system problems, sleep disorders, depression, anxiety, job dissatisfaction and burnout [48–52].

8. Economic aspects of occupational health and safety

Industrialization and technological developments lead to occupational injuries and environmental risks which cause socioeconomic losses not only in Turkey but also in various countries including European countries [53, 54]. Even though there are precautions taken against occupational risks in developed European countries, occupational accidents and diseases due to profession have not been prevented for long years. This indicates that it is hard to overcome occupational disease and accidents. On the other hand, the gradual decrease in the occupational accidents and diseases related to profession in European countries in the past 10 years show us the importance of the application of the precautions. In this regard, when we consider the outcomes of our country, we should support the regulations in the workplace, create the incentive systems, ensure effectiveness of Occupational Health and Safety Workers' Representation and Occupational Health and Security Committees, and increase the trainings in order to foster social consciousness instead of waiting the employees and employers to apply these rules [53]. According to ILO, it has been reported that there is 1.25 trillion dollars loss each year due to the OHS problems [55]. In Turkey, the loss of only social security systems has been reported as approximately 4 million Turkish Liras per year [56]. According to the report of Turkey Statistical Institute (TSI) 2007 Gross Domestic Product values, the total costs of occupational accidents in Turkey are almost 35 billion Turkish Liras per year [57].

9. Occupational health and safety services for health professionals in Turkey

The Central Council of the Turkish Medical Association initiated first studies about the health status of health professionals in Turkey in 1989 with the slogan entitled "This is our health." Then, a database on the subject was aimed to be established [58]. The requirement of a committee, which should carry out the studies regarding the health and safety of employees working in the hospital, was discussed in 1999–2000, and the pilot studies were initiated, but these studies were not completed. The precautions were compiled intended for the safety of the health professionals in the 15th item of the "Notification on Principles and Procedures related to Ensuring the Patient and Staff Safety and Protection in Health Institutions" which was approved in 29th of April, 2009. In a sense, this document created a basis for the establishment of the related OHS unit [59, 60].

The concept of OHS services is defined by the USA Occupational Safety and Health Administration (OSHA) and National Institute of Occupational Safety and Health (NIOSH) and Hospital Safety Committee (HSC). Accordingly, they are explained as first step health care which improves and protects the health of health professionals and fulfills the therapeutic care services for outpatients. An effective hospital occupational health program must include at least the following titles: recruitment examinations including a full medical history, periodic inspections, health and safety trainings, immunizations, health counseling, environmental control and surveillance, record systems of health and safety, and coordinated planning between hospital departments and services. Furthermore, the environmental control and surveillance program should be a part of the occupational health program, and required precautions should be taken by performing risk analyses [61–63].

The Association of Public Health Professionals (originally HASUDER in Turkey) Occupational Health Working Group (OHWG) performed their first activity in Gazi University, Medical

Faculty, Ankara, hosted by the Chief of Medicine in 18th of September, 2010. The title of the activity was "Health Organization Workshop of Health Professionals in Hospitals." In the meeting organized by HASUDER and OHWG in 2010, the problems of hospitals' OHS units were discussed. It was stated that the basic problem was financing. Since it was not written in the Health Communication Notification (HCN), it was decided that employers should pay the costs in case the payment is a burden to employees. It has been emphasized that the committees in hospitals still need a consensus about organization, function and finance [64].

Public health professionals had opportunity to take advantage of occupational health and safety services with the help of "Regulation on the Provision of Patient and Staff Safety" published by the Ministry of Health in 6th of April, 2011, in Official Gazette and the "Occupational Health and Safety Law" published by Ministry of Labour and Social Security in 30th of June 2012 in Official Gazette [62]. Later, hospitals were informed about the implementation of regulations and establishment of the employee safety unit via instructions issued in 14th of May, 2012 [63]. In the HASUDER-OHWG workshop hosted by Antalya Akdeniz University, Medical Faculty, between 13th and 15th of September, 2012, organization of employee health units, functions of these units and financing topics were discussed and a model was established by receiving the contributions and recommendations of the participants. In this workshop, the aim was to recommend a model about the organization of OHS units which are newly established or to be established in the future, functions of these units and financing. It was also aimed to notify the institutions and create a common language in terms of naming of the concepts in order not to complicate the issues [65].

10. Conclusion and recommendations

Even though there are occupational health and safety units in state and private hospitals, personnels and service are still insufficient. Thus, the applications can vary from region to region, institution to institution as well as person to person [61]. In this regard, as in various European countries, it can also be efficient in Turkey in case the government controls the services for health professionals and if these services are provided independent of the requests of individuals [63]. The health of the health professionals also affects the health of the community. When we consider the regional differences of hospitals (such as technical equipments, staff and quality of services), which as an emerging economy, it is very important to apply the regulations that are specified by national laws and procedures across the country including the public and private health sector. Additionally, it is also crucial to decrease the exposure to occupational risks of health workers [66].

Occupational health and safety is very crucial in terms of the resource allocation. In case a portion of the economic resources can be allocated to OHS trainings and organization, the efficiency and increments in the production due to the application of OHS regulations can accelerate the economic growth and the development [67]. The importance of OHS precautions has increased due to the understanding that the costs that eliminate the harm of occupational accidents are higher than the costs that prevent the occupational accidents [68].

11. Protection from risks in a workplace for health professionals

The things which must be done at workplace:

1. The employment of necessary staff (doctors, work safety experts and other technical staff) must be made paying attention to the legal rules.
2. The health security council at work must be formed and the continuity of its meetings must be gained.
3. The sources of danger at work must be mentioned.
4. The evaluation of risks at work must be made.
5. When making risk evaluation, the problems, the staff who are responsible and the solution suggestions must be mentioned with their dates.
6. The urgent situation plans and practices at work must be made.
7. The medical examinations of the staff when they start their jobs and when they return their jobs must be filed.
8. They must be given work health and security education.
9. Everything must be recorded [69].

Funding

No outside funding was received for this study.

Conflict of interest

The authors declare that they have no conflict of interest.

Acknowledgement

We would like to thank The Association Of Public Health Professionals (originally HASUDER in Turkey) Occupational Health Working Group (OHWG) for whom we have used the documents and the information shared by the researchers who are the source of the article.

Author details

Nilgun Ulutasdemir[1*] and Ferdi Tanir[2]

*Address all correspondence to: nulutasdemir@yahoo.com

1 Faculty of Health Sciences, Avrasya University, Trabzon, Turkey

2 Cukurova University School of Medicine, Adana, Turkey

References

[1] Mollaoğlu M, Kars Fertelli T, Özkan Tuncay F. Assesment of perceptionrelatingworkenvironment of nursesworking in hospital. Journal of FıratHealth Services. 2010;5(12):17–30.

[2] Ulutasdemir N, Cirpan M, Oztürk Copur E, Tanir F. Occupational risks of health professionals in Turkey as an emerging economy. Annals of Global Health. 2015;81(4):522–529.

[3] İzgia MC, Öztürk Türkmen H. Worker health and occupational safety status determination of subcontractor health workers in Akdeniz University. Turkish Journal of Public Health. 2012;10(3):160–173.

[4] Azap A, Ergönül Ö, Memikoğlu KO, Yeşilkaya A, Altunsoy A, Bozkurt GY, Tekeli E. Occupational exposure to blood and body fluids among health care workers in Ankara, Turkey. American Journal of Infection Control. 2005;33(1):48–52.

[5] CDC. Health Care Workers. Available from: http://www.cdc.gov/niosh/topics/healthcare/ [21.07.2016].

[6] Türkay M, Aydoğdu E. Prevalence of occupational accidents in health workers working in a university hospital and factors affecting accidents.Ankara: 3rd National Congress of Health of Health Workers; 18-20 November 2011. Oral proceding-20.

[7] Sabuncu H.H. Organization of work safety, occupational health and environmental health services in hospitals. Available from: http://www.saglikcalisanisagligi.org/tezler2/Yeni.pdf [Accessed: 07 July 2016].

[8] Meydanlıoğlu A. Health and safety of health care workers. Journal of Balıkesir Health Sciences. 2013;2(3):192–199.

[9] Özkan Ö, Emiroğlu O.N. Occupational health and safety services towards hospital health employees. Journal of Cukurova University School of Nursing. 2006;10(3):43–51.

[10] Beşer A. Management health risks and of health workers. Journal of Dokuz Eylül University School of Nursing Electronics. 2012;5(1):39–44.

[11] Toraman AR, Battal F, Ozturk K, Akcin B. Sharps injury prevention for hospital workers. International Journal of Occupational Safety and Ergonomics (JOSE). 2011;17(4):455–461.

[12] Occupational risks of health workers (First Edition). Ankara: Publications of the Turkish Medical Union. October 2008. Available from: http://www.ttb.org.tr/kütüphane/sc_meslek_riskleri.pdf [Accessed: 21 July 2016].

[13] Emiroğlu C. Occupational risks in health sector and legal arrangements. Turkish Medical Union Journal of Occupational Health and Safety (JOHS). January-February-March 2012;43:16–25.

[14] Özgür N. In terms of radiological risks situation in Turkey and ways of protection. Ankara: 4th National Congress of Health of Health Workers; 16-17 November 2013. pp. 79–87.

[15] Özen S, Tosun P, Helhel DS. Electromagnetic field levels in emergency call centers and assessment of occupational exposure. Istanbul: Symposium on Electromagnetic Fields and Effects 2011, Safety for Environment and Public Health; 2011. pp. 221–224.

[16] Vehid S, Ethem Erginöz E, Eray Yurtseven E, Çetin E, Köksal S, Kaypmaz A. Noise level of hospital environment. TAF Prev Med Bull. 2011;10(4):409–414.

[17] Beyzadeoğlu H, Cengiz İ. Health advocates and risks of health workers. Journal of Health Thinking and Medical Culture. Autumn 2013;28:28–33.

[18] Baykam N. Prevention of ventilation in the control of Hospital Infections. Hospital Infection Control Handbook. Scientific Medicine Publisher. Hospital Infections Association Publication no: 2. Ankara; 2004. pp. 373–381.

[19] Öcek Z. Chemical hazards and risks, consequences and precautions of health workers in the working life (Group Work-3). Ankara: 3rd National Congress of Health of Health Workers; 18-20 November 2011. pp. 170–172.

[20] American Society of Health-System Pharmacist. ASHP guidelines on handling hazardous drugs. 2006;1172–1193.

[21] Occupational Safety and Health Administration (OSHA). Technical manual. Section VI, Chapter 2. 1999.

[22] Occupational Safety and Health Administration. Controlling occupational exposure to hazardous drugs. Directorate of Technical Support. OSHA Instruction TED 1.15. 2001.

[23] Connor T H, Mcdiarmid M A. Preventing occupational exposures to antineoplastic drugs in health care settings. Cancer Journal of Clinicians. 2006;56:354–365.

[24] Ergönül Ö. In terms of Blood and body fluids the situation in Turkey and ways of protection. Ankara: 4th National Congress of Health of Health Workers; 16-17 November 2013. pp. 68–74.

[25] Evyapan F. In terms of Respiratory infections the situation in Turkey and ways of protection. Ankara: 4th National Congress of Health of Health Workers; 16-17 November 2013. pp. 75–78.

[26] Demir M, Tuncay E, Yentürk E, Kanmaz D. Risk of tuberculosis infection in employees of chest diseases hospital. Anatolian Journal of Clinical Investigation. 2014;8(2):57–61.

[27] Bloodborne infectious diseases: HIV/AIDS, Hepatitis B, Hepatitis C; Overview. Center for disease control and prevention. CDC 24,7 [Electronic Journal]. Available from: http://www.cdc.gov/niosh/topics/bbp/[Accessed: 2016-07-22].

[28] Himmelreich H, Rabenau HF, Rindermann M, Stephan C, Bickel M, Marzi I, Wicker S. The management of needlestick injuries. Deutsches Arzteblatt International. 2013;110(5):61–67.

[29] İlhan MN, Durukan E, Aras E, Türkçüoğlu S, Aygün R. Long working hours increase the risk of sharp and needlestick injury in nurses: The Need for New Policy Implication. Journal of Advanced Nursing. 2006;56(5):563–568.

[30] Ortabağ T, Güleşen A, Yava A, Bakır B. Exploring the frequency of sharps injuries and affecting factors among health care workers in a university hospital. Anatolian Journal of Clinical Investigation. 2009;3(4):208–212.

[31] Vaizoğlu S, Güler Ç. Ergonomic hazards and risks, consequences and precautions in the health workers working life (Group Work-4). Ankara: 3rd National Congress of Health of Health Workers; 18-20 November 2011. pp. 177–183.

[32] Ergonomics, US Department of Labor Occupational Safety and Health Administration (OSHA). Available from: http://www.osha.gov/SLTC/ergonomics/[Accessed: 2016-07-23].

[33] Ilhan M.N. The removal of the electromagnetic field map in a medical faculty hospital and the health status of employees in areas with electromagnetic fields. [Doctoral Thesis]. Ankara: Ankara University Health Sciences Institute; 2007.

[34] Kevser Tarı, Selçuk K, Çevik C, Karataş G, Mercan Y. Prevalence of past 12 months waist pain and associated factors in midwives and nurses working in a maternity and pediatric hospital in Istanbul. Ankara: 3rd National Congress of Health of Health Workers; 18-20 November 2011. pp. 134–135.

[35] Aksakal N, Ilhan M. N, Yüksel H, Kurtcebe Ö, Bumin M.A. The frequency of back pain and factors affecting nurses health officer and patient caregivers working in a university hospital. Turkish Medical Union Journal of Occupational Health and Safety (JOHS). 2009;32:38–46.

[36] Al B, Zengin S, Deryal Y, Gökçen C, Arı Yılmaz D, Yıldırım C. Increased violence towards health care staff. JAEM. 2012;11:115–124.

[37] Roche M, Diers D, Duffield C, Catling-Paull C. Violence toward nurses, the work environment and patient outcomes. Journal of Nursing Scholarship. 2010;42:13–22.

[38] Hahn S, Hantikainen V, Needham I, Kok G, Dassen T, Halfens R.J. Patient and visitor violence in the general hospital, occurrence, staff ıntervention sand consequences: a cross-sectional survey. Journal of Advanced Nursing. 2012;68(12):2685–2699.

[39] Kingma M. Workplace violence in the health care sector: a problem of epidemic proportion. International Nursing Review. 2001;48:129–130.

[40] Nau J, Halfens R, Needham I, Dassen T. The de-escalating aggressive behaviour scale: development and psychometric testing. Journal of Advanced Nursing. 2009;65:1956–1964.

[41] Chen W.C, Hwu H.G, Kung S.M, Chiu H.J, Wang J.D. Prevalence and determinants of workplace violence of health care workers in a psychiatric hospital in Taiwan. Journal of Occupational Health. 2008;50:288–293.

[42] Violence. Occupational hazards in hospitals. NIOSH. Available from: http://www.cdc.gov/niosh/docs/2002-101/pdfs/2002-101.pdf [Accessed: 2016–07-21].

[43] Ayranci U, Yenilmez C, Balci Y, Kaptanoglu C. Identification of violence in turkish health care settings. J Interpers Violence. 2006;21:276–96.

[44] Franz S, Zeh A, Schablon A, Kuhnert S, Nienhaus A. Aggression and violence against health care workers in Germany, a cross sectional retrospective survey. BMC Health Services Research. 2010;10:51.

[45] Stathopoulou HG. Violence and aggression towards health care professionals. Health Science Journal. Available from: http://www.Hsj.Gr. [21.07.2016].

[46] Estryn-Behar M, Van der Heijden B, Camerino D, Fry C, Le Nezet O, Conway P.M, Hasselhorn H.M, NEXT Study Group. Violence risks in nursing-results from the European 'next' study. Occupational Medicine (Lond). 2008;58(2):107–114.

[47] Pınar T, Pınar G. Sağlık çalışanları ve işyerinde şiddet. TAF Preventive Medicine Bulletin. 2012;12(3):315–326.

[48] Delbel J.C. De-Escalating workplace aggression. Nursing Management. 2003;34,30–34.

[49] Yiğitbaş Ç, Deveci S.E. Mobbing towards health workers. Turkish Medical Union Journal of Occupational Health and Safety (JOHS). 2011;42:23–28.

[50] Teymourzadeh E, Rashidian A, Arab M, Akbari-Sari A, Hakimzadeh S.M. Nurses exposure to workplace violence in a large teaching hospital in Iran. International Journal of Health Policy and Management. 2014;3(6):301–305.

[51] Hutchinson M, Wilkes L, Jackson D, Vickers MH. Integrating individual, work group and organizational factors: testing a multidimensional model of bullying in the nursing workplace. Journal of Nursing Management. 2010;18(2):173–81.

[52] Efe S.Y, Ayaz S. Mobbing against nurses in the workplace in Turkey. International Nursing Review. 2010;57(3):328–34.

[53] Yılmaz F. Occupational health and safety in European Union and Turkey: Measuring the effectiveness level of occupational health and safety boards in Turkey. [Doctoral Thesis]. Istanbul: Istanbul University Social Sciences Institute; 2009.

[54] State Planning Organization (SPO). Specialized commission report on labor market. SPO: 258-ÖİK, 564. Ankara; 2001. p. 86.

[55] International Labour Office (ILO). http://laborsta.ilo.org/cgi–bin/brokerv8.exe Available from: [Accessed: 2016-07-23].

[56] Social Security Institution (SSI). Statistical report. Ankara; 2006.

[57] The Ministry of Labor and Social Security, "T.C. National job health and safety policy document II (Draft Text); 2009-2013:3.

[58] Akdeniz University Faculty of Medicine-The Association of Public Health Professionals (HASUDER), Occupational Health Working Group (OHWG). Health of health workers workshop: organization, functions and financing of employee health units. Available from: http://www.saglikcalisanisagligi.org/dosyalar/SCS-Calistay-Antalya.pdf [Accessed: 23 July 2016].

[59] Health of Health Workers Working Group. Agenda. Health of Health Workers. 2000;1:1–4.

[60] Türk M. Occupational health and work safety services in health institutions in Turkey. Ankara: 3rd National Congress of Health of Health Workers; 18-20 November 2011. pp. 98–103.

[61] Official newspaper. Notification hazard classes related to occupational health and safety. Available from: http://www.resimigazete.gov.tr/eskiler/2009/11/20091125-10.htm [Accessed: 23 July 2016].

[62] T.C. Ministry of Health. Regulation for the provision of patient and employee safety. Available from: http://www.resimigazete.gov.tr/eskiler/2011/04/20110406-3.htm [Accessed: 23 July 2016].

[63] T.C. Ministry of Health. Employee security policy. Available from: http://www.saglik.gov.tr/TR/belge/1-15642/calisan-guvenligi-genelgesi.html [Accessed: 2016-07-23].

[64] The Association of Public Health Professionals (HASUDER) Occupational Health Working Group (OHWG). Health organization of health workers in hospitals workshop report. Ankara; 2010.

[65] The Association of Public Health Professionals (HASUDER), Occupational Health Working Group (OHWG). Health of health workers workshop report. Ankara; 2012.

[66] Ünal D, Aycan N. Examination of sociodemographic characteristics and reasons for application to the workplace health unit in Manisa MAY Textile Industry workers. Journal of Nursing Forum. 2003;6(1):49–57.

[67] Cilengiroglu O. Occupational health and safety in Turkey during the EU adjustment process.[Masters Thesis]. Izmir: DEU Institute of Social Sciences; 2006.

[68] Erkul I. Business and social security law. Eskisehir: Anadolu University Publication No:79; 1986. pp. 183.

[69] Health of Health Workers Working Group. Ogan H. (Preparer to Broadcast).Worker health and safety for health workers. First edition. Ankara; October 2014. pp. 48.

Chapter 4

HSE Management for a Sound Work Environment: Strategies for Improving Health Safety and Environmental Indicators Through Ergonomic Design Thinking

Marcello Silva e Santos,
Maria da Conceição Vinciprova Fonseca,
Marcelo Marcio Soares,
Bernardo Bastos da Fonseca,
Maria Victoria Cabrera Aguilera and
Ananda Halfeld Alves Fernandes

Additional information is available at the end of the chapter

http://dx.doi.org/10.5772/66650

Abstract

Ergonomic Design Thinking (EDT) is a project management methodology that takes advantage of two important concepts or themes in carrying out project actions. The first is Design Thinking itself, a project management approach originally proposed by Tim Brown, who knew beforehand the full potential of design tools, techniques and maybe we should add idiosyncrasies. Designers have "their own way" of following through and carrying out issues such as deadlines and sequences, for example. This logic is similar to another important theme: ergonomics. The main objective of ergonomics is adapting work systems to workers themselves. By doing so, its professionals dig deep into the social technical fabric of a workplace and use recurrent and iterative strategies in order to search for a perfect fit for a given workstation. EDT as a modeling guide for workspace projects have been used in Brazil for quite some time. This text outlines an interesting experience in which EDT was used as a conception tool in building a new health safety and environmental (HSE) management system model for construction sites. A real case–an ongoing construction work–was used to contextualize the experiment and better define the various instruments of this HSE model. Due to the work environment and predominant job characteristics available, the EDT approach did quite well in terms of serving its project management purpose, as it was confirmed when the new system became fully functional.

Keywords: ergonomics, health, safety and environmental, management systems

1. Introduction

Ergonomic Design Thinking (EDT) was originally conceived as a project management model used for both organizational design and project management. Thus, it must be understood as an effective tool for project management optimization in general applications, not only design projects. The main idea is the general concept that makes possible the application of creative reasoning and intensive end-user participation in the course of project management just like designers do when solving problems in their during the creative work process. Conventional planning and decision making methods flow in a linear manner, way different from creative thinking. "HSE," short for health safety and environmental management is a fairly new–and crucial–area in operations management [1], especially for those organizations that fall in the category of complex systems, which require a 360° approach to operations management. HSE decision analysis in such environments is definitely a challenge, since prioritizing becomes difficult due to multiple criteria to assess, risks to ponder and level of severity in terms of collective impact and workstation evaluation.

This chapter outlines the general model for carrying out HSE project management approaches in the workplace. In practical terms, it adds up to existing management systems that are designed to comply with industry and government standards. In order to contextualize the theoretical framework, a case study is employed as a guide for the implementation of a specific HSE Management system for the construction industry. In order to achieve that, the work team used a set of strategies set forth by the EDT modeling, as proposed by Santos and Soares [2]. The choice of this industry segment is due to the fact this particular sector is known by the lack of qualification of its workers. As a consequence, it is acknowledged worldwide as a low performance sector in terms of controlling its operational and environmental risks and hazards.

The EDT approach used in the course of the process being described in here was carried out to help the consolidation of a HSE management system for building construction sites. Thus, the complex nature of the work process is dictated not by its operational characteristics or the usual determinants of larger complex systems (i.e., oil refineries, nuclear power plants, medical emergency rooms, construction sites), but because of the difficulties posed by management issues, such as control and enforcement of safety and health practices in this type of work. The building and construction industry struggles to set up and maintain effective action plans and indicators when it comes to health safety and environmental processes, commonly referred to as HSE. In fact, its workers ranks the second most exposed to work accidents of all industrial segments in Brazil [3], just behind the transportation sector.

The consequence of poor control and management of HSE in construction sites leads to a series of problems ranging from plain inconvenience of mishandling or misinterpretation of data to more serious ones such as poor HSE plans and management systems. The end result of this is converted into an unpleasant workplace, health and safety issues and all sorts of environmental problems. Without the continuous optimization of HSE processes, it becomes virtually impossible to positively improve the work conditions for the workers and all those directly or indirectly affected by the outcomes of bad planning, poor management and actual

http://dx.doi.org/10.5772/66650

operational actions. The EDT methodology provides the "real work element" for such continuous improvement since it is more palatable and credible for containing and merging the expectations of the workers with the needs of the productive process.

2. Theoretical framework

The concept of EDT derives from the general Design Thinking concept, originally proposed by Brown [4], who defined it as "a discipline that uses the designer's sensibility and methods to match people's needs with what is technologically feasible and what a viable business strategy can convert into customer value and market opportunity." However, the concept has proven to adhere well beyond the design and marketing spectrum. Likewise, ergonomics is not only a tool for improving work conditions but also an effective way for a better design of products and for bringing productivity, therefore economic advantage, for the organizations [5, 6].

The idea behind Design Thinking has been the appropriation of creative thinking and intuitive response actions involved in the design process in a wider range of situations, outside the Design field. With that in mind, Santos and Soares [2], proposed a combination of the Design Thinking "creative based process management" and Ergonomics for Project Management processes in which the interaction between users and those in charge of designing work environments or systems is key to an overall conformity to job design principles and user needs, as well as to the compliance to normative standards and labor legislation (**Figure 1**).

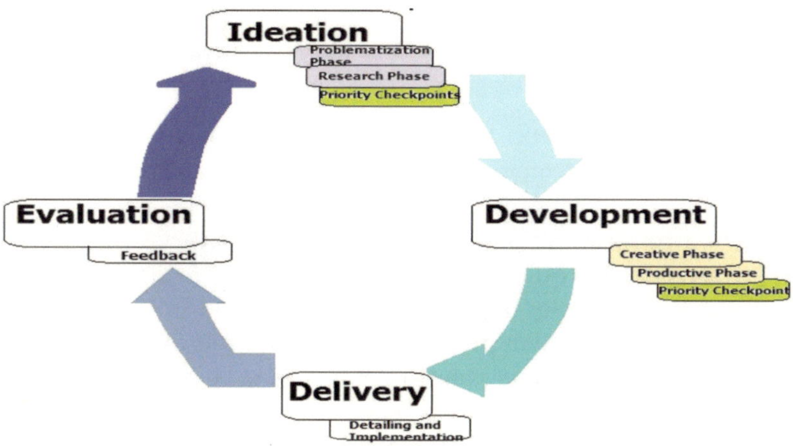

Figure 1. General EDT model.

The concept of Ergonomic Design Thinking, hereafter called by its acronym EDT, adheres to the notion of employing formal creative thinking built from within the spectrum of real work actors. Santos et al. [7] points out that nobody knows better about the job than workers themselves. Therefore, it would seem counterproductive not to take in consideration their insights when trying to improve their work conditions, workstations or work systems. Thus, Ergonomic Design Thinking must be seen as a general model for participatory actions in the workplace, which in turns make areas such as Ergonomic Design, product development and, definitely, job design all possible terrain for its dissemination.

All those actions demand sequential steps somewhat flexible to allow creative thinking to flourish. In fact, the model started to be used in setting up HSE management systems in a variety of organizations. Health, safety and even environmental processes are all connected to an important participatory demarche in operational work activities. They should not be dealt with in an insulated manner, without proper integration. Our lives and our health can be affected by poor design choices [8]. By inference, this is true for HSE interventions. Regardless all constraints that may arise in workplace, engineers are mostly responsible for all work environment inadequacies. It is fair to say that companies intuitively use Design Thinking to an extent, even without realizing it. However, organizations resist in taking a human-centered approach because they cannot grasp at the perspective of trying something entirely new. Something that arises out of their control, in the midst of a balance of users' needs, technology and organizational constraints.

The building construction industry is characterized by particularities not present in any other industry. First off, it is entirely a project driven industry, no "final product" is ever alike, due to geographic and geological variations and differences in systems and components (building materials and techniques). Contrary to regular transformation industry, storage and logistics is not a major direct problem, but quality control is as important as in regular manufacturing. Then, we have the construction site issue. It functions as a temporary factory, one in which often times workers share not only tasks but also sleeping rooms with fellow workers. Yet, there is no true attachment to the workplace. It will be used and discharged at the end of every project.

Likewise, teams are not permanent, rarely replicated in the next construction projects. For instance, if a work activity is identified as critical under HSE standpoint, a training program is immediately implemented. However, those trainees will hardly experiment the same constraints in their next construction project. In fact, sometimes they may be even out of the industry all together, since the Building Construction Industry (BCI) is also characterized as a temporary job for many people, employing large amounts of unskilled workers. In order to be successful, a HSE qualification program must take in consideration all of these issues and address them properly. HSE plans and systems need to be pragmatic, safety driven and easily assimilated by everyone. Although every construction site has its own characteristics, it is possible to establish a replicable model that could pass on the essence of prevalent work characteristics that are present in such work environments.

Brazilian Standard NBR 12284 [9] defines a construction site as "a set of areas destined to the execution and support of construction industry Works, divided in operational and living areas." In other words, it is the entire portion of land in which a construction plan will be materialized, plus the storage areas, equipment and machines (concrete mixers, cutters, welding pits, etc.) and also sleeping barracks, restrooms and lunchrooms for the workers. They are inherently hazardous places to work and to "live" and often times workers have no choice but to stay overnight since construction works and construction sites may be located in remote areas.

The risk of developing musculoskeletal disorders is very high because in most cases there is no possibility of mechanical material handling. Besides heavy lifting, workers are often exposed to awkward postures, making the BCI the leading industry in ergonomics-related risks and impacts [10]. This is all aggravated by heat stress (especially in some states) and other environmental hazards. Making matters worse, the exposure to job design constraints, issues such as lack of job security and crew rotation, increases the chances of accidents and infirmities [11].

In places with a strong public welfare system, occupational data and epidemiological statistics are usually more precise and credible than in wealthier nations in where private modalities of welfare are more common. In Brazil, the National Institute for Social Security needs a robust database to be able to control multiple benefits it manages, from retirement pensions to temporary disability payments, including indemnity for work accidents and work-related illnesses and disorders. The system operates in an intricate network of information, trying to avoid errors and frauds by individuals and companies, since resources are becoming gradually scarce as population grows older and people live longer.

The cost of liability insurance for example is not a flat rate for generic or specific business categories. It is based on each tax identification number, so that companies with poor occupational health and safety indicators will pay gradually more if they keep neglecting this particular issue. In fact, there is a curiosity that comes from epidemiological data in places with more harsh labor and welfare legislation. More precise data in occupational health and safety data lead to a false perception of low standards. Because it tends to increase reporting of the negative outcomes of HSE management systems, it ends up making them look ineffective when compared to other countries. Often times, countries that are known for having poor health and safety standards may look better in some of those standards. It happens because reporting of occupational health and work accidents in those countries are also very poor.

Table 1 contains data extracted and combined from various Brazilian government agencies [12, 13], and illustrates epidemiology severity data for the BCI, in terms of frequency.

According to the social security annual statistics report [13], the impact in terms of occupational health and safety figures is exponentially more relevant once the data are further detailed. Out of more than 700.000 work-related accidents, about 121.000 of those involved victims under 29 years old and the building and construction industry is responsible by nearly one third of those figures. Soares [14] sums up the economic impact of work accidents saying that social security agencies alone spends around US$ 1 billion in wages, benefits and other expenditures, whereas the private sector spends another US$ four to five billion due to poor work conditions.

Number of work related accidents in Brazil			
Year	**Accidents**	**Illnesses**	**Deaths**
2000	363.868	19.605	3.094
2005	499.680	33.096	2.766
2010	709.474	17.177	2.753
2011	720.629	16.839	2.038
2012	705.239	14.955	2.731

Table 1. Number of work related accidents in Brazil.

By looking at the data in **Table 1**, it is tempting to infer that the decrease in the number of fatalities and illnesses–despite an increase in number of accidents–is due to the improvement of work conditions and incentives in preventive measures, such as the use of personal protective equipment (PPE) and collective protective systems. However, it is fair to say there has been an overall raise in awareness by workers and consciousness by organizations to employ systematic measures for the improvement of work conditions as a whole, which includes some human factors and ergonomics actions. Another Brazilian government-funded agency, SESI [15], points out a list of prevailing occupational illnesses and injuries in the building construction industry. **Table 2** lists the main causes and preventive measures for each one of those work hazards.

Main occupational illnesses in the building construction industry		
Occupational illnesses	**Causes**	**Preventive measures**
Hearing loss	Long exposure to noise above 85 dB	PPE usage, collective acustic protection
RSI/WMD	Repetitive tasks for long periods of time	Adequate use of rest, work breaks, warm up and stretching exercises
Silicosis and asbestosis	Inhalation of fine particles from asbestos and ceramic like materials	Use of appropriate PPE (masks)
Air embolism	Underwater work or incorrect use of equipment	Compliance with norms and decompression procedures
Low back pain	Inadequate material handling/load lifting	Avoidance of heavy lifting/use proper mechanical aid
Heat stress	Long exposure to ultraviolet sun rays	Proper hydration and avoiding excessive sun exposure
Contact dermatitis	Contact with cement, chemicals or even certain types of PPE	Use PPE appropriately, clothing, gloves and so on

Table 2. Prevailing occupational illnesses in the BCI.

The last letter in the HSE acronym represents the environmental aspects of the workplace, its aspects and impacts for the people and surroundings. The BCI is by large the biggest generator of solid urban waste. Considering the entire production chain, the environmental impact varies according to the construction system (steel, wood, concrete and combinations) but it is always high nonetheless. The graph on **Figure 2** shows the distribution of greenhouse gas emissions in Canada per each type of economic sector, placing the BCI, negatively, as a major contributor for another important environmental indicator. The full report makes considerations on the data gathering methodology that raises the numbers for the construction sector even more by adding indirect impact of the entire production chain [16].

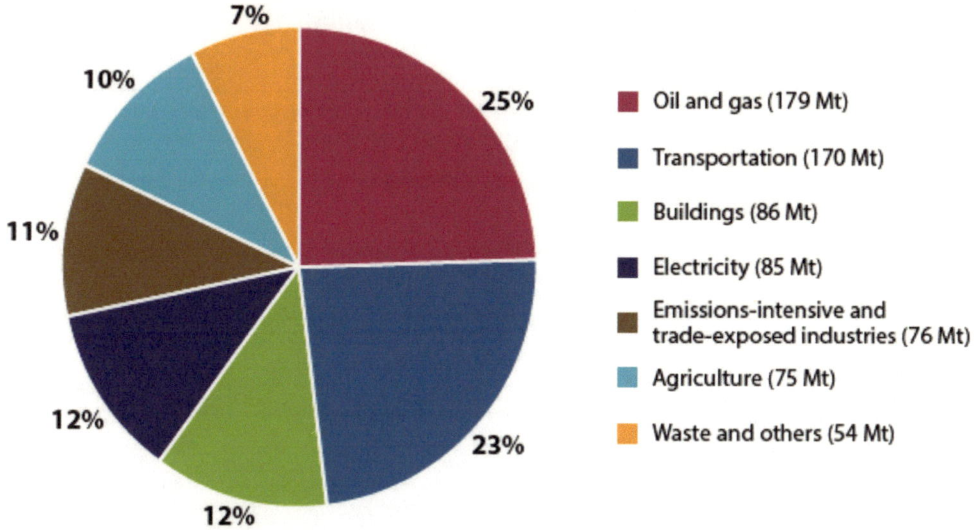

Figure 2. Distribution of greenhouse gas emissions in Canada.

The BCI's overall environmental impact has to do with processes such as mining (for sand and aggregates), which accounts for great levels of energy consumption, transportation (mostly by diesel trucks) and last, but not least, generation and inadequate disposal of debris and unused materials. In fact, a great deal of the problem is created by the lack of understanding and acknowledge on the part of authorities and society in general regarding the negative impacts of clandestine disposal and the benefits of solid urban waste disposal. Usually, BCI solid waste is considered a minor environmental hazard, which impact is basically due to the volume generated. However, this misconception hides some by-products of those residues, such as the proliferation of undesirable species as rats, cockroaches and insects that may act as vectors for various diseases [17].

3. Developments and building of the model

Every management system model is embedded with particularities and peculiarities related to whatever industry it will be applied in. As obvious as it sounds, it is not rare to find various management systems that are built with generic mechanisms and inferences that will not adhere to specific industry processes. HSE management systems for transformation industries, for example, follow certain criteria based on the nature of their operations. If it involves air pollution, preventive actions related to chimneys are necessary to be considered in maintenance plans. Likewise, occupational health issues will vary immensely, depending on the presence of certain risk factors, prevalent climate and atmosphere, geographical and geological conditions and other aspects. Using EDT methodology, the team involved in a consulting project in charge of carrying out an analysis for setting up a new HSE program in a construction site, decided to start with a brainstorm activity with all key actors in a construction site. The resulting master plan for the program establishment and project implementation was defined according to an operational sequence based on **Table 3**.

Analysis phase	- Observational analysis–during this step the consulting project team performs workstation and activity analysis to assess program guidelines and best practices opportunities
	- Tool applications–methodological tools are applied to facilitate the process and maximize results
	- Project reporting–it represents the communication of the results to the various agents (managers, engineers and workers)
Validation phase	- Project validation–the analysis report needs to be verified and validated by all stakeholders, in search of errors and misinterpretations
	- Complementation–as a consequence of the validation, new analysis, as well as the application of different tools, as required to assess different issues regarding environmental impacts, health and safety hazards
	- Final check–before implementation the report is consolidated and a master action plan for the implementation is set forth. Elements for the banner launch are discussed and a schedule is set for that and other implementation actions
Implementation phase	- Capacitation training sessions–as part of the HSE management system, workers and other people involved in the various construction site processes need to be informed about and qualified to deal with risks hazards and to adhere to best practices related to health safety and environmental issues
	- Banner launch–it is a symbolic event in which a banner illustrating best practices for HSE in construction sites is presented to workers and all the people using this workspace. The banner is displayed in conspicuous places inside the area, such as the lunchroom

Table 3. Construction site HSE management system model.

4. Case study

In order to contextualize the proposed HSE management model, it is necessary to apply the various mechanisms embedded in that model in a real scenario. Thus, resourcing to a case study strategy will not only allow linking theoretical aspects to real life issues but also facilitate the understanding of the model outreach. A construction site in a city in the State of Rio de Janeiro, Brazil, was sorted out in this particular case, for better illustrating the application of the HSE management model being described herein. Starting from a general plan outlined in **Table 1**, each step will be now described in their most relevant aspects.

4.1. Analysis phase

An essential tool for the alignment of different perceptions and to establish, by consensus or inferences, guidelines for an appropriate diagnostic of a work process is the brainstorming tool. It is a group creativity technique that tries to lock up on conclusions for a specific problem by gathering ideas spontaneously, as they are verbalized by its members. It was extremely useful for the definition of the various analysis instruments, methods and techniques that were eventually implemented to carry out the work process analysis in the construction site. The methodology for the analysis phase follows this sequence:

(a) Global analysis: It starts with a walkthrough, a broad analysis that consists in screening both organizational and operational aspects that may influence or be influenced by the work environment. General characteristics of the enterprise need to be assessed and reported during this phase.

(b) Observational analysis: The observational analysis is concurrent to steps c and d. It displays the characterization of a given work activity or workstation with a short description of aspects and impacts, as well as images of eventual improvements derived from the action. Some activities were contextualized through a technique called "Animated Simulation" [18], in which workers simulate through role playing the real activity that takes place.

(c) OIT check list: It is dynamic questionnaire (excel spreadsheet) based on the most relevant aspects of the ergonomics check list from the International Labor Organization (OIT in Portuguese). It is built in a way that automatically displays essential information on the task and serves for prioritizing the HSE action plan.

(d) Conversational analysis: During the process of applying the HSE check list, the analysts use a technique called "Conversational Action" [19], which helps to "extract" important information more naturally from the workers.

(e) Tallying and validation of data: The results of all different actions, especially the application of action tools, must be tallied and validated.

(f) Reporting (communication) of findings: The last phase in the HSE management system methodology is reporting and communication. In ergonomic actions, there is a preliminary report, called hot report that is elaborated during the process in order not to miss important information that might be otherwise lost if left out for the end of the field work.

In order to stay focused on the main aspects of the analytical phase, a detailed explanation of all the different steps of the entire methodology could be counterproductive and cause some misconceptions. Therefore, only essential operational actions will be detailed, in other words, further description is concentrated only on those actions that were performed by the HSE analysis team in the course of the field work, prior to diagnostics itself.

4.1.1. Observational analysis

In order to better illustrate the context in which the observational analysis is carried out, an excerpt of the HSE appreciation report is needed. The selected operation consists of a concrete pouring process, for building foundation.

Foundation concrete : works

Workstation: bricklayers and helpers

Situation: The activity consists the pouring of concrete into wooden forms that were previously built and laid in each appropriate frame mold (**Figure 3**). The concrete is pumped up into the forms by a flexible pipe that connects to a special concrete truck, which has a rotating mixing barrel to maintain concrete's chemical characteristics and pumping equipment (**Figure 4**). It may be performed under any climatic and atmospheric condition, except in case of heavy, persistent rain.

Figure 3. Foundation pouring.

Figure 4. Vibration of concrete mix.

Visible impacts: Awkward postures of workers, unsafe and unstable surfaces used as base for the activity, evidence of poor planning of activities and lack of operational and safety

http://dx.doi.org/10.5772/66650

training, harsh environmental conditions without proper protection, no incentives for hydration, soil contamination, and improper solid waste disposal.

Checklist application: According to the HSE checklist, the activity stands in 53% of adequacy, which is considered acceptable by HSE standards. However, there are several small interventions that may raise current levels to a "good" mark, which will be listed in the following section.

Improvement opportunities: (a) adding collective protection equipment (CPE) in the workplace, (b) specific training designed to fit the nature of each task, (c) adding schedule pauses for certain activities, (d) potable water available next to the operations, (e) improving signage and notes in conspicuous places, (f) allocate appropriate areas for solid waste disposal, avoiding soil contamination, and (g) ladders and scaffoldings should be used as standard equipment.

Environmental aspects: During preparation and pouring of concrete at least 5% of material was lost (**Figure 5**), either by poor connection between hose joints, spills from truck equipment or by lack of care of workers as pumping is carried out. A lot of the concrete mix flows over the wooden forms or under them straight to the ground (**Figure 6**). In this particular case, not only the mix cannot be re inserted into the frames but also part of mix percolates through the soil.

Figure 5. Concrete waste.

Figure 6. Wood waste.

After hardening, the left overs could be reused as aggregates. However, the chosen destination for discharge in our context was a rented dumpster.

Regulatory and legal framework: (a) Environmental–according to environmental regulations (CONAMA 307 Resolution, Art. 3rd) concrete and wood are class A materials, which mean they should be fully recycled. Therefore, it is not legal to discharge them as trash. If they are not used in the site, they must be taken to special facilities for future reuse; **(b) health**–in terms of occupational health, this work activity does not comply to several aspects of labor regulations, especially offending Articles in NR 15 (safety risks), NR 17 (ergonomics) and NR 24 (cleaning and conservation. Occupational hygiene standards are also disrespected, especially those related to temperature and weather exposure; **(c) safety**–there are some issues with regard to PPE and Collective Protective Equipment (CPE) during operations, as established by NR 6, NR 7 and NR 9, as well as Brazilian Technical Standards (ABNT/ISO).

4.1.2. OIT check list

According to the check list, which is a quite simple excel spreadsheet that runs in a tablet device, the particular activity shown in here stands in 53% of adequacy. This is considered acceptable by HSE standards. The filling of the check list is pretty much a straightforward operation. The analysts basically use their fingers to a scroll down list with two possibilities: yes and no/not applicable. The annotations in red are "manually" registered by the analysts and contain important notes for diagnostics and the establishment of "HSE Program Improvement Opportunities," which is incorporated in the resulting Program's Best Practices Banner (**Figure 7**).

SMS	PROGRAMA DE GESTÃO DE SMS
SMS	Análise de SMS do Trabalho - Setor ou Função: Diversos

Lista de verificação para mapeamento de Ergonomia, Segurança, Saúde e Sustentabilidade
Situação : **CONCRETAGEM DA FUNDAÇÃO**

Questões Ergonomia e Segurança do Trabalho:		
1	Esforço muscular estático?	NÃO
2	Repetitividade notória na atividade?	NÃO
3	Reporte espontâneo de sobresforço?	NÃO
4	Situações envolvendo torção ou flexão de tronco?	SIM
5	Situações envolvendo torção ou flexão de cabeça?	SIM
6	Situações envolvendo extensão de tronco ou cabeça?	SIM
7	Adução ou abdução de membros inferiores?	NÃO
8	Adução ou Abdução de membros superiores?	SIM
9	Bancadas de trabalho com altura incompatível?	NA/NO
10	Utilizam-se ferramentas perfuro-cortantes?	NÃO
11	Utilizam-se ferramentas vibratórias?	SIM
12	Pouco espaço de trabalho ou mobilidade insuficiente?	SIM
13	Faltam proteções ou barreiras contra incidentes e acidentes no local de trabalho	SIM
14	Comunicação com colegas insatisfatória ou insuficiente?	NÃO
15	EPI's incompatíveis com a natureza do trabalho?	NÃO
16	As pausas pré-programadas são inexistentes ou reduzidas?	SIM
17	Existe movimentação de carga acima de 23 KG?	NÃO
18	Espaço de guardar de ferramentas de trabalho inadequado?	NA/NO
19	Falta monitoramento dos operários?	NÃO
20	Existe movimentação manual de carga de grandes dimensões?	NÃO

Questões Associadas a Saúde, Eficácia e Sustentabilidade:		
21	Água potável indisponível ou quantidade insuficiente?	SIM
22	Deslocamentos são frequentes ao longo do dia?	NÃO
23	O trabalho ocorre em turnos?	NÃO
24	Existe trabalho externo com temperaturas acima de 36°?	SIM
25	Existe trabalho externo com temperaturas abaixo de 10°?	NÃO
26	Legibilidade de documentos/avisos/sinalização insatisfatória?	SIM
27	Iluminação é incompatível com a natureza da atividade?	NÃO
28	A temperatura é inadequada para realização normal da atividade?	NÃO
29	O nível sonoro encontra-se fora dos limites de tolerância?	NÃO
30	Falta proteção contra intempéries no local de trabalho?	SIM
31	Falta treinamento específico para a natureza dos serviços?	SIM
32	Falta liberdade ou autonomia na realização das tarefas?	NÃO
33	Sudorese excessiva ou aparentemente deficitária?	NÃO
34	Urina muito clara ou muito escura?	NA/NO
35	Faltam orientações básicas operacionais *por escrito* para novos colaboradores	SIM
36	Descarte de materiais e lixo não atende princípios ambientais (seleção e reaproveitamento)	SIM
37	A escolha dos insumos segue apenas o critério econômico-financeiro	SIM
38	As técnicas construtivas são ecologicamente ineficazes	SIM
39	O esgoto de alojamentos/cantina/escritório de obra despeja direto na rede ou fossa	SIM
40	A etapa da obra gera poluição ambiental (gases, fumos, poeiras em quantidade)	SIM

91 a 100% = Excelente
71 a 90% = Boa
51 a 70% = Aceitável
31 a 50% = Moderada
Menos de 31% = Necessita Intervenção Imedi

Computar somente iten: aplicáveis e percentual proporcional.

Quantidade de itens favoráveis:

Percentual de itens favoráveis:

Total de respostas: 53%

Percentual Condição Ergonômica:

Observações: (em vermelho são exemplos de regi
Foi dada orientação postural e de ajustes em equipamentos/dispositivos de trabalho.
Recomendada modificação de procedimento.
Recomendada pausas sem obrigatoriedade de intervalos determinados.
Recomendada modificação de procedimento.
Recomendada alteração/adoção de EPI (botas, luvas, capacete, calça e camisa)
Recomendado rodízio de funções.

Figure 7. Tube underpinnings concrete pouring checklist.

http://dx.doi.org/10.5772/66650

4.1.3. Tallying and validation of data

The assessment of HSE conditions report (A-HSE report) is the final product of the analytical process. It is submitted to the scrutiny of all participants in the process during the validation phase that follows. The data submitted are organized according to relevance and synthetized in tables and diagrams, a selection of which is presented as follows. Some data come from images; therefore, they were not translated into English.

(a) Severity table for each analyzed work activity: It presents the full list of work processes according to the degree of adequacy in terms of HSE standards (**Table 4**). The lowest levels are color coded in red, moderate levels in orange and the highest ones in yellow.

Work Activities	Occupational Severity Levels
Pipe foundation excavation	**15**
Rebar positioning	**40**
Formwork (floors)	**45**
Columns spacing bars	**45**
Foundation Rebars	**48**
Concrete work (floors)	**50**
Formwork (Pilars and Beams)	**50**
Rebar work (Pillars and Beams)	**50**
Walls elevation	**53**
Foundation Concrete Work	**53**
General formwork	**55**
Plastering	**55**
Painting	**55**
Flooring	**60**

Table 4. Occupational severity levels.

(b) SIC diagram: SIC stands for sum of critical indices (**Table 5**). It is elaborated using an excel spreadsheet with embedded logic formulas and simple macros. It produces a clear picture of the overall work system and allows prioritizing of an eventual HSE action plan according to the risks and severity aspects and impacts present in each activity, following gravity, trend and urgency of action. Its color coding allows instant visualization for

CL OIT	G	U	T	EN	SIC
73%	3	3	3	5	185

Resultado da Adequação do Posto Administrativo - Canteiro de Obras REDUC		Legenda
91 a 100% = Excelente	73%	
71 a 90% = Boa		<100
51 a 70% = Aceitável		100,1-150
31 a 50% = Moderada		150,1-275
Menos de 31% = Necessita Intervenção Imediata		>275,1

Table 5. Sum of critical index for one particular workstation.

sorting out those work activities that require urgent or immediate action (red and purple, respectively). The percentage result indicates solely the severity level, therefore, it is possible for an operation (or workstation) to have high adequacy to HSE standards but require special attention, just like in the example provided as follows.

(c) Final overall portrait of the work system (RFO): As the name indicates, it is a final picture of the entire operation, as assessed by the analysis team. As in every HSE management system context, the picture is a representation of a current situation and it may be modified by several work environment circumstances and, of course, improved by appropriate HSE actions (**Figure 8**).

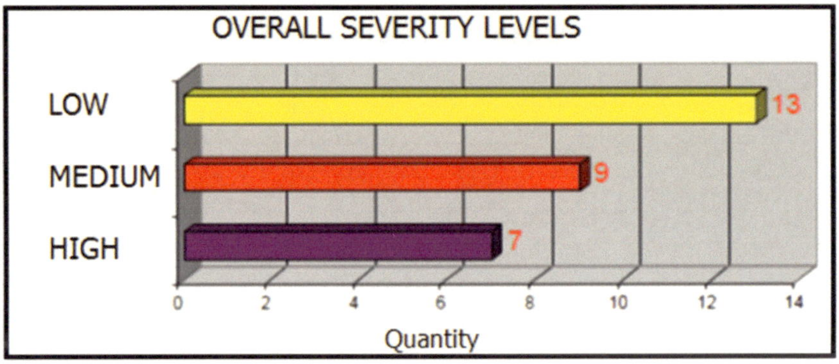

Figure 8. Overall severity levels for the work environment.

4.2. Validation phase

A final and validated A-HSE report version, along with the HSE best practices banner is a combined final product of the validation process. They are both now ready to be implemented, along with its subsequent HSE program guidelines, which become part of a broader HSE management system or policy. The HSE program will be consolidated during the implementation phase that follows.

4.3. Implementation phase

Possibly, the best illustration of what Ergonomic Design Thinking can produce to the work systems is the final result generated in the process of application of our proposed HSE management system model. The PROSMS (HSE program) best practices banner is a quite symbolic representation of a collective interaction of people and disciplines to contribute towards a better, safe and sound work environment (**Figure 9**). It is "launched" in an event that combines training and social gathering. The banner is displayed in a conspicuous place inside the construction site, usually the lunchroom.

Figure 9. HSE best practices banner.

5. Final considerations

The health safety and environmental management system model shown in this chapter resulted from the application of a methodology called Ergonomic Design Thinking. It is based in two equivalent project management approaches easily found in design and ergonomics projects. Preliminary experiences had already shown that the Design Thinking approach proved itself effective not only for design project actions [20]. Using the general concept derived by Tim Brown's original approach, EDT adds the basic principle of ergonomics: No one knows better the work performed than workers themselves. And this is particularly true for ergonomics projects, since ergonomists and human factors professionals know exactly what is necessary to allow a perfect synchronicity between their logic and knowledge and the correct listening and interpretation of what some have been calling "the voices of the shop floor" [21].

Naturally, much more real-world experiences are needed to master the technique and to affirm without shadow of a doubt that this model is more effective and others. It is definitely better than the ones the authors have previously utilized. Hopefully, the publication of this text will provide an opportunity to disseminate the method and its tools to a more broad and global audience. To wrap up the discussion, a list of important issues that should be addressed before employing any participatory action method is presented down below.

1. Know who the end users are (and involve them),
2. Employ one methodology (not "the" methodology),
3. Emphasize iterativeness (not only interactivity),
4. Break the linear ruling (but don't run in circles),
5. Don't ever give up (make lemonades with sour lemons),
6. Be trustworthy (never hide flaws and mistakes),
7. Communication is key, fear control is out (stimulate reports),
8. Simulate real job situations (don't guess what you don't see),
9. Never fear opposition (get opposition closer),
10. Present well viewpoints (perception is more important than reality).

One last issue is important to address. The best thing about both EDT methodology and our proposed HSE management system model is that they do not employ intricate, complex tools or instruments. It is completely "open source", so there is no proprietary software to buy, copyrighted materials or any similar exclusivity resource. Even some known and common tools used during ergonomic assessments, like the Corlett Diagram and the NIOSH equation for horizontal load handling, for example, are not under any trademark imposition. It was designed as a collective experiment in which all free men is more than welcome to join in.

Author details

Marcello Silva e Santos[1]*, Maria da Conceição Vinciprova Fonseca[2], Marcelo Marcio Soares[3], Bernardo Bastos da Fonseca[4], Maria Victoria Cabrera Aguilera[4] and Ananda Halfeld Alves Fernandes[5]

*Address all correspondence to: marcellosantos@hotmail.com

1 UniFOA–University Center of Volta Redonda, UGB–Geraldo di Biase University, Volta Redonda, Brazil

2 UniFOA–University Center of Volta Redonda, Volta Redonda, Brazil

3 UFPE–Federal University of Pernambuco, Recife, Brazil

4 UERJ–Rio de Janeiro State University, Resende, Brazil

5 UGB–Geraldo di Biase University, Valença, Brazil

References

[1] Kleindorfer, P. R., Singhal, K., Van Wassenhove, L. N., (2015). Sustainable Operations Management. Production and Operations Management. 2005; 14: 482–492. doi:10.1111/j.1937-5956.2005.tb00235.x

[2] Santos, M., Soares, M., (2016). Ergonomic Design Thinking – A Project Management model for Workplace Design, In: Ergonomics in Design: Methods and Techniques, Soares, Marcelo & Rebelo, Francisco (Eds.), Boca Raton: CRC Press.

[3] Ministry of Labor and Employment (MTE), (2015). National Strategy for the Reduction of Work Accidents 2015–2016, Brasília – Brasil.

[4] Brown, T., (2008). Design Thinking, Harvard Business Review, 86: 84–92.

[5] Jonh R. W., (2014). Fundamentals of systems ergonomics/human factors. Applied Ergonomics. 2014; 45(1): 5–13. DOI: 10.1016/j.apergo.2013.03.021

[6] Hendrick, H. & Kleiner, B. (Eds.), (2016). Macroergonomics: Theory, Methods, and Applications, Boca Raton: CRC Press.

[7] Santos, M. et al., (2009). Can We Really Opt in Terms of Ergonomic Methodologies and/or Approaches? In: Annals of XVII IEA Congress, Beijing.

[8] Santos, M., (2012). The PhOCoe Model – Ergonomic pattern mapping in participatory design processes. Work (Reading, MA). 2012; 41: 2643–2650.

[9] Brazilian Association of Technical Standards, (1991). Areas of experience in construction sites - ABNT / NBR 12284, 1991.

[10] Haslam, R. A. et al., (2005). Contributing Factors in Construction Accidents. Applied Ergonomics. 2005; 36(4): 401–415. doi:10.1016/j.apergo.2004.12.002

[11] Couto, H., (1995). Ergonomics Applied to Work: Technical Manual of the Human Machine, Vol I and II, Belo Horizonte: ERGO.

[12] Fundacentro[CS627], (2012). Jorge Duprat Figueiredo Foundation for Safety and Work Medicine. 2012. Annual Report on Safety and Prevention at Work (in Portuguese), available at http://www.fundacentro.gov.br/ 05/23/2016.

[13] INSS, National Institute of Social Security (Brazil), (2013). Statistical Data Annual Report, available at http://tabnet.datasus.gov.br/tabdata/cadernos/cadernosmap.htm 05/27/2016

[14] Soares, L., (2008). The financial impacts of labor accidents on the Brazilian budget: A political and pedagogical alternative to reduce expenses, Dissertation ISC, Brasília, 2008.

[15] SESI - Social Services Industry, Safety and Health in the Construction Industry in Brazil - Diagnosis and Recommendations for the Prevention of Accidents at Work, SESI Annual Report (in Portuguese), 2015, available at http://arquivos.portaldaindustria.com. Br / app / conteudo_18 / 2014/07/28/6973 / ESTUDODESST2015.pdf? R = 0.625125239368 05/27/2016

[16] Environmental Canada, (2013). Canada's Emissions Trends Report, available at https://www.ec.gc.ca/ges-ghg/985F05FB-4744-4269-8C1A-D443F8A86814/1001-Canada's%20Emissions%20Trends%202013_e.pdf

[17] Karpinsk, L.A. et al., (2009). Differentiated Management of Civil Construction Waste: An Environmental Approach. Porto Alegre: Edipucrs, 2009.

[18] Santos, M., Vidal, M., (1999). The Use of Live Animated Simulations for Malfunction Prevention in Facilities, In: Proceedings of the I Latin American Ergonomics Conference", Salvador, Brazil, September, 1999.

[19] Bonfatti, R. J., Vidal, M. C., (2003). Conversational Action – An Ergonomics Approach to Interaction[I8], In: Rethinking Communicative Interaction, Grant, P., (Ed.), John Benjamin Publishing Company, Philadelphia, US ,pp. 108–120.

[20] Siqueira, O. A. G., Cunha, L. S., Pena, R. D. S. F., Corrêa, B. S., Amorim, M. E., (2013). Project Methodology for Design, Ergonomic Methodology and Design Thinking: Convergence in Developing Solutions in Design. In: Cadernos Unifoa Especial Design - Centro Universitário de Volta Redonda Fundação Oswaldo Aranha, Vol. 1, December, pp. 49-66.

[21] Santos, M., Vidal, M., (2008). Success or Failure, What a difference a good plan makes: A Benchmarking Case Study of Robotics Implementation in Two Public Pharmacy Facilities. In: ICIEOM (International Congress on Industrial Engineering and Operation Management), Rio de Janeiro.

Chapter 5

Health-Promoting Leadership Culture and its Role in Workplace Health Promotion

Simona Šarotar Žižek, Matjaž Mulej and
Vesna Čančer

Additional information is available at the end of the chapter

http://dx.doi.org/10.5772/65672

Abstract

Purpose: The law on health and safety at work was implemented in Slovenia in 2011. On this basis, Slovenian organizations started preparation and implementation of workplace health promotion (WHP) programs. The article reports on research of the Slovenian leaders' leadership style concerning their employees' health following the new legislation.

Methodology/approach: We conducted a survey on health-promoting leadership culture (HPLC) in Slovenian organizations in 2013. As the collected data were measured on an ordinal scale, the median measured the central tendency, the frequency tables were displayed and the nonparametric independent samples Mann-Whitney U test was used.

Findings: There are no significant differences between of leaders and employees. leaders and employees. They have positive relations with each other.

Research implications: For health-promoting leadership culture (HPLC) of leaders, it is important that the criticism is appraised; every opinion is shared fairly; one can identify with the values of the organization; everybody has the necessary latitude, nobody has to work against their personal values; in conversations, employees experience their acceptance like they are. These elements can be the basis for developing new leadership's styles, which can determine employee's health through health promotion's activities.

Practical implications: Findings support new requisitely holistic approach to managing health-promotion and its role in WHP.

Originality/value of paper: The available literature offers no similar concept for implementing health promotion on workplace.

Keywords: Ethics, Health, Health-promoting leadership culture, Management, Leadership, social responsibility, workplace health promotion, work environment

1. Introduction

The implementation of the new legislation on health and safety on work place [1] aims to insure employee's health and safety. The reported-about research investigates the current situation of the leadership culture, especially concerning the health promotion. Therefore, we first focus on the current research state concerning the role of health and safety at workplace and especially on the attitudes about health promotion and leaders' behavior.

Workplace environment covers a crucial part of human life. Therefore, workplaces should be designed in a human-oriented way, following the tradition since 1950s. Having healthy, satisfied and motivated employees, who enjoy health and safety at workplace, enables positive economic outcomes, increased competitiveness and profitability; it also affects other factors in the business positively. Healthy jobs and workplaces benefit workers/employees, customers and shareholders, citizens, society and the state [2, 3].

After EU-OSHA [4], workplace health promotion (WHP) encompasses employers', employees' and society's actions improving the human health and well-being at work. It integrated the following activities in the defined area:

- Improving the way work is organized;
- Improving the work environment;
- Encouraging employees' involvement in health-related activities;
- Encouraging personal development.

These activities must be planned and implemented. The key factors therefore include leaders who can form, stimulate and change the work environment substantially. This special part can present the health-promoting leadership culture (HPLC), which we measured in Slovenian organizations. One should know this culture as the basis to develop special trainings for improving leaders' leadership style.

The law on health and safety at work was implemented in Slovenia in 2011. On this basis, Slovenian organizations started their workplace health promotion (WPC) programs. The responsibility for WPC is on employers and leaders, i.e., managers on all hierarchical level. For successful implementation of WHP, leaders should lead their employees in a way, supporting their health. The reported-about research covered the Slovenian leaders' leadership style concerning their employees' health after implementing the said new legislation. We conducted a survey on health-promoting leadership culture in Slovenian organizations in 2013, which is the first research, conducted after the above-mentioned new legislation passing.

2. Health and workplace health promotion

2.1. Health

Health is fundamental in every person's life. To understand health well, we will first quote some general definitions:

The World Health Organization (WHO) defined in 1947 the concept of health as a state of complete physical, mental and social well-being and not merely as the absence of disease or infirmity [5]. WHO argues that health care is not only an individual concern but also a concern for the whole society; it is both a private and a social good. Therefore, the environment in which one lives crucially impacts one's health.

Gabrijelčič Blenkuš et al. [6] (summarized after [7]) suggested that good health of the population matters for poverty reduction, economic growth and long-term development of society. There are four mechanisms for individuals' contributions to the economy in high-income countries: labor productivity, labor market participation, education and savings and investments. Good health increases the likelihood of participation in the labor market, whereas poor health reduces earnings and increases retirement [6, pp. 17–18].

O'Donnell [8] states that health promotion is important because it helps people to discover synergies between their passions and optimal health in order to give them more motivation and support to change their lifestyle. It is about achieving the necessary balance between the physical, emotional, social, spiritual and intellectual health. A combination of learning experiences facilitates changes of lifestyle: it enhances awareness, increases motivation, builds skills, and, most important, creates opportunities to access environments making positive health practices one's easiest choice [8].

2.2. Health and safety at workplace

Business/production processes can cause accidents in the working process; therefore, safety and health at work matter. It covers protecting the safety, health and welfare of people engaged in work.

The European Agency for Safety and Health's Information Bulletin, entitled "The Business Benefits of good occupational Safety and Health" [9], finds organizations aware that they can reduce the number of accidents and occupational diseases with ensuring the health and safety at work.

Health and safety at work impact the organizations' business performance visible inside them. Surveys show that every euro invested in WHP brings the company 2.5–4.8 € by diminishing costs due to absenteeism [10]. Presentism could be added: it diminishes productivity of the ill, but present workers/employees.

Indirect cost for support to early retirees could also be added. So could repairing of equipment that is wrongly used due to poor work of the ill, but present workers/employees. WHP can often prevent these and similar troubles; outcomes can be estimated with opportunity cost methods.

Protection of life, health and work ability of workers is a fundamental task of safety and health at work. Detecting all types and levels of burdens and harms of occupational risks and evaluating how serious the work and health risks are, decisively impacts both on business and employees. Assuring the health of employees is a multidisciplinary activity, which may include physiology, psychology of work, labor pathology and industrial toxicology, assessment of work ability, ecology of work, the epidemiology, ergonomics and working environment, social medicine, hygiene, the legal aspect of comprehensive care staff and others [11, p. 5]. Even more: it should be interdisciplinary to include synergies of the listed activities and attain the requisite holism of approach and requisite wholeness of outcomes of interventions [12].

2.3. Workplace health promotion

Promotion of health and safety is a planned combination of educational, political, environmental, organizational and values/culture/ethics/norms-related mechanisms that support measures and conditions of lives of individuals, groups and communities.

WHP complements occupational safety and health measures as parts of the combined efforts of employers, workers and (inter-)national authorities to improve the health and well-being of humans at work. Its definition reads: "the combined efforts of employers, employees and society to improve the health and well-being of people at work" [13]. It belongs to the most prominent approaches, improving the individual well-being at work, encouraging health-related individual and organizational learning, and creating health-sustaining work environments (for example, see [13]).

WHP programs support prevention efforts, classified as primary (directed at employees that are generally healthy), secondary (directed at individuals already at high risk because of certain lifestyle practices or abnormal biometric values) and tertiary (referred to as disease management, directed at individuals with existing ailments such as asthma, diabetes, cardiovascular disease, cancers, musculo-skeletal disorders and depression; their aim covers ameliorating the disease or retarding its progression) [14].

This (requisitely) holistic concept incorporates, first, the improvement of the work organization and the working environment; second, the promotion of active participation of all stakeholders in the process; plus third, the encouragement of personal development [13].

It affects the psychological and mental areas, social sphere, facilitating prevention activities and general cognitive and affective outcomes [15, p. 41].

In Slovenia, the concept of the workplace-health promoting is relatively recent. It resulted from the concept of WHP, which has significantly evolved since becoming prominent in the 1970s. Then, several companies in some European countries showed great enthusiasm in start-up projects to humanize the world of work; but they quickly abandoned these endeavors which poorly matched corporate (short-term and narrowly defined?) interests. In the late 1980s, workplace health and safety mattered again, primarily due to the European Commission's Framework Directive on health and safety [16]. It reorientated occupational safety and health in Europe [17].

Interest in WHP has grown considerably at the national and supra-national levels in Europe over the past 5 years. This trend can be related to the far-reaching economic and social changes that are taking place in the EU member states due, in part, to increasing international competition [18].

Companies can improve their safety and health at work with their own activities. If they adopt and further promote the legislation of safety and health at work, they improve work safety (and reduce their visible and opportunity costs, too).

In recent years, the attention to ensuring safety and health at work in the so-called green workplaces has grown. European Union is trying to coordinate economic growth with the need to protect the environment. To reach this, one shall lower the emissions of greenhouse gasses, raise energetic efficiency, stimulate renewable sources of energy and lower the amount of waste produced. This was the basis for creating green workplaces, encouraging humans towards preserving and restoring their environment. Implementing of these goals can improve safety and well-being of workers too [19, p. 5].

EU wants to protect the workers of all its member states against threatening dangers at work in every workplace.

In Slovenia, the area of promotion of health and safety at work is legally defined. In 2011, the amendments to the "Act on occupational safety and health" added the area of health promotion and management of psychosocial risks in the workplace. ZVZD-1 to the square of 6 and Article 32 stipulates employers to plan and implement health promotion in the workplace, and to provide the necessary means and method of monitoring its implementation [20, pp. 32, 53]. Evidently, the law is being realized [21].

2.4. WHP models and approaches

Nowadays, there are several different models for implementing WHP. Even though health promotion programs are becoming more wide-spread, they are just programs, mainly focusing on individual health behaviors rather than working conditions. Reducing employee health risk requires innovations in job design, workplace culture, organizational systems and management practices [22].

The most common approach introduces the health circles. They help occupational safety and health (OSH) professionals and leaders in private enterprises, public organizations and authorities. Health circles can be implemented by dividing the process into six successive phases [22]: first, prerequisites (consensus, project group); second, preparation (needs analysis, expert discussion); third, practiced interventions based on needs analysis; fourth, presentation of results to project group and staff; fifth, implementation of measures; sixth, evaluation by circle participants, staff and management, including company documents.

The "management circle" enables continual improvement. Relating to a circle should demonstrate that the end of one activity leads to the beginning of another activity. Health management therefore becomes a permanent organizational process. The implementation of the health circles lasts about 15 months.

These efforts can be called contributions to socially responsible business and enjoy support from social responsibility.

3. Social responsibility

In ISO 26000 [23], corporate social responsibility (CSR) is defined as one's responsibility for one's impacts on society. It addresses human values, culture, ethics and norms (VCEN); hence, ISO 26000 is advisory rather than obligatory. Namely, it is not organizations as legal entities that decide, but their authorized decisive members and stakeholders, WHP included. CSR is necessary as the way out from the current global socioeconomic crisis [24], hence for purely social and economic reasons: responsible behavior is cheaper and more efficient/effective than a one-sided and short-term one. The one-sided and short-term behavior is namely the common denominator of the causes of the current crisis: under the label of a totally free market, monopolization destroyed the market as the room of equality of all its participants (for details see e.g. [25]).

In ISO 26000 [23], all seven contents are linked by two notions from systems theory:

1. Interdependence (replacing ethics of independence, e.g., "the boss is the only decisive person and benefits," and ethics of dependence, e.g., "subordinates only obey orders and may be irresponsible"); and
2. Holistic approach (replacing ethics of one-sidedness without interdisciplinary creative cooperation linking mutually different viewpoints for synergetic insight and action).
3. VCEN is equally crucial as knowledge ("if somebody knows how to shoot with a gun, the choice depends on VCEN whether the person will shoot upon human or upon a paper target, rather than on knowledge").

Before passing of the ISO 26000 by which the entire world unified the official definition of the CSR, there were several and quite diverse definitions of CSR from different authors such as Esposito [26], EU [27], McWilliams and Siegel [28], Waldman et al. [29], Campbell [30],… We can look at CSR as a contemporary version of informal systemic behavior designed to attain requisite holism of behavior of influential persons and enterprises [31].

Need for CSR should be integrated into enterprise policy as the enterprise's basic, general and long-term orientation that arises from organization's vision [32]. Therefore, principles of stakeholder interests, enterprise developmental, economic and social (SIEDES) responsible (enterprise) policy matter, WHP included. See **Table 1**.

The principles of SIEDES responsible (enterprise) policy
1. To reach the enterprise's business excellence and hence to find its way out from its crisis, responsible enterprise policy should stress the regular innovating in its policy
2. Arising from VCEN innovations, expressed as persons interests, the enterprise's general definitions of its policy depend on interests of its important stakeholders
3. Enterprise stakeholders should be conscious about their long-term well-being in interdependence with all life species
4. All of them should once again rethink their long-term interests (benefits)
5. Their (process, product/services, information, etc.) consequences
6. Their willingness/ability to innovate them toward their responsible, requisitely holistic behavior concerning all other humans (families, co-workers, other citizens and planet Earth residents, life forms and nature—with predominating of long-term interests concerning all of them)
7. Enterprise's developmental orientation (e.g., exploitation of opportunities of its internal and external environmental development)
8. Their economic orientation (with striving for politics of economic responsibility toward all inhabitants of the world), and
9. Their social orientation (toward ecological, socially and else-how responsible goals and social desire consideration, also toward social community) should also be innovated all the time
Source: Šarotar Žižek et al. [32].

Table 1. The principles of stakeholder interests, enterprise developmental, economic and social (SIEDES) responsible (enterprise) policy.

To implement the SIEDES, responsible (enterprise) leaders must innovate their VCEN and VCEN of organization's stakeholders. This tackles leadership and ethics. People find ethics inapplicable to the real world; they assume that ethics covers short and simple rules like "Do not lie," "Do not steal" and "Do not kill" [33]. Ethics is specified in principles of ISO 26000 [23, p. 11]: A behavior of employees in an organization should be based on the values of honesty, equity and integrity, which imply a concern for people, animals and the environment and a commitment to address the impact of its activities and decisions on interests of stakeholders. They make the difference between the one-sidedly commanding boss and the cooperative leader, who uses ethics of interdependence to attain requisite holism of approach and requisite wholeness of outcomes of his/her organization by social responsibility, including ethics (as defined above).

4. Models of leadership and business ethics

Let us elaborate the third principle of ISO 26000, defining ethics, in the case of business leaders. We exposed leadership, because leadership support (leaders' involvement in, and promotion of, activities, policies and practices that encourage the development of social responsibility and related climate) was identified as an essential component of successful WHP programs [34, p. 1]. In leadership, there are some normative models or leadership theories. A normative model or leadership theory consists of explicit moral norms for analyzing leaders and leadership; the first such normative model is servant leadership: leaders are supposed to serve followers [35, p. xxvii]. Test of servant leaders is whether the people they serve become better, freer, healthier, and more likely to serve others [35, p. xxviii]. Another, transformational leadership is based on the idea that leaders and followers must improve each other's moral.

Transformational leadership develops from conflict: dialogue emerges from conflicts in which both leaders and followers move toward agreement about their shared moral values. In this process, change comes when both side agree on higher order [i.e., less one-sided and short-term, N.B. authors] values [35, p. xxviii].

Such models prove authentic leadership. It is a process that draws from positive psychological capacities and a highly developed organizational context, which results in greater self-awareness and self-regulated positive behaviors on the part of leaders and employees, fostering positive self-development [36, p. 243]. Therefore, this leadership focuses on ways leaders' self-knowledge contributes to making leaders effective and ethical. In authentic leadership, morality results from a leader's quality of authenticity [35, p. xxviii]. Waldman et al. [29, p. 1718] confirmed connection between transformational leadership and CSR.

Responsible leadership is a values-based and through ethical principles-driven relationship between leaders and stakeholders. They are connected through a shared sense of meaning and purpose through which they raise one another to higher levels of awareness, motivation and commitment for achieving sustainable values creation and social change [37]. Theory of responsible leadership holds leaders responsible for a wider range of people and things. The main issues include leaders' moral obligations covering their stakeholders and interests inside and outside their organizations.

Spiritual leadership is described as occurring when a person in a leadership position embodies spiritual values such as integrity, honesty and humility, creating the self as a person of someone who can be trusted, relied upon, and admired. Spiritual leadership is also demonstrated through behavior, whether in individual/employees reflective practice or in the ethical, compassionate, and respectful treatment of employees [38, p. 663]. On the other hand, the spiritual leadership comprises the values, attitudes and behaviors that are necessary to intrinsically motivate humans' self and others providing their sense of spiritual survival through calling and membership [39, p. 711].

Ethical leadership demonstrates the normatively appropriate conduct through personal actions and interpersonal relationships and promotes such conduct to followers through two-way processes, e.g., communication, reinforcement and decision-making. Moral variables, such as honesty, trust, fairness, openness and consideration, are connected to ethical leadership [40, p. 120].

From the viewpoint of this research on leaders' social responsibility (including ethics, of course) concerning WHP, ethical leadership matters because it is [41, pp. 601–608] positively related to:

- Ability to identify a proximate, ethical role model during one's career
- An ethical context supporting ethical conduct
- Agreeableness
- Conscientiousness
- Level of leaders' moral reasoning
- Followers' ethical decision-making
- Prosocial behavior
- Followers' satisfaction, motivation, organizational commitment

And negatively related to:

- Neuroticism and
- Machiavellianism

Ethical leaders must develop their reputation for ethical leadership. Employees' and key external stakeholders' perception and reality of managers' reputation at all levels matter for ethical leadership [42]. Reputation for ethical leadership is based on two essential pillars: perception that the potential leader is both a moral person and a moral manager. A moral manager is a role model for ethical conduct, communicating regularly about ethics and values, and using rewarding system to hold everyone accountable to the values and standards [42, p. 141]. One must develop one's reputation for ethical leadership. Therefore, leaders must be consistent and proactive about incorporating ethics into their leadership agenda, to match principles of social responsibility, including WHP.

5. Health-promoting leadership culture

Leading and business culture have an important effect on the psychological and physical health of employees. In order to provide permanent high quality performance, it is important that leaders are trained regarding models of health exposure and strain [43]. Jiménez et al. [44] states that managers can learn a new concept of "healthy leadership" with organized specific workshops and seminars, which allows them the realization of this concept and helps them to change their behavior in the workplace. This concept requires a style of human and relationship-oriented focus of subjective well-being of employees and managers.

Health-promoting leadership, by definition, creates a culture for health-promoting workplaces and values (VCEN), inspires and motivates employee participation in this development. It is a critical part of the organizational capacity for health promotion, including managerial knowledge and skills practiced as organizational policies and structures, supporting a health-promoting workplace. Therefore, leaders' involvement in the systematic and requisitely holistic development of both the physical and psychosocial work environment matters [45].

Leaders impact safety and health behavior of employees in two fields. First, leaders shape organizational process and management sub-systems; thus, they control various health-related physical and psycho-sociological characteristics of the work and working environment. Second, leaders influence employees' safety behavior and health through day-to-day direct and personal interaction and communication [46].

Health-promoting leadership culture combines various leadership styles and behaviors to design healthy workplaces [47].

Leaders can create workplaces that enhance health, if they can recognize and reduce mismatches between employees and their organization. Mismatches can occur in six areas of work life [48]: workload, control, reward, community, fairness and values.

Let us finish the summary of the theoretical bases of research that we report about here.

6. Hypotheses development

WHP is aimed to have physically and mentally healthy employees. Employees should gain also their psychological well-being, including positive (social) relations with others [49]. One's well-being depends on warm, rewarding and trustworthy relationships with other people, concerns about the well-being of others, the ability of empathy, affection and intimacy, and understanding the giving and receiving components of human relationships. These attributes can be developed with WPH: the Law on Safety and Health at Work [1] exposes that leaders should develop and implement each activity for generating safety and health at work in cooperation with employees. Such relations could be developed through:

- Provision of relevant information to employees and the preparation of the communication strategy;
- Involvement of employees in the decision-making process;
- Preparation of the work culture based on partnership;
- Organization of work tasks and processes in terms of the positive impact on health; and
- Introduction and implementation of policies and practices, which enhance employee relations.

The first hypothesis therefore considers positive (social) relations with others:

H1: Leaders and employees established positive (social) relations with others.

Furthermore, the Law on Safety and Health at Work [1] says that leaders must ensure health and safety at work, hence prepare, plan and implement activities, and consult with employees about the mentioned activities, including informing employees about activities. On the other hand, employees must:

- Respect and implement measures to ensure their safety and health at work;

http://dx.doi.org/10.5772/65672

- Perform their duties with optimal care, to protect their and other persons' lives and health;
- Use the funds for the work, safety devices and personal protective equipment in accordance with their purpose and the employer's instructions; carefully handle them, and make sure their perfect condition.

Leaders must build circumstances for healthy workplace and culture in the organization. Hence, the second hypothesis reads:

H2: There are no statistically significant differences between leaders' and employees' health-promoting leadership culture in their working environment.

7. Empirical research–methodology

7.1. Measuring instrument

We used the questionnaire entitled health-promoting leadership culture, developed by Jiménez et al. [50]. It can detect attitudes and behavior regarding health-promoting leadership culture with seven dimensions: health awareness, low workload, control, reward, community, fairness and values. There are two versions: a self-assessment version for leaders, and an external assessment version for employees. Thus, leaders can evaluate their own style of healthy leadership, while employees can evaluate the extent of healthy leadership culture of their superiors. The questionnaire HPLC has 44 items, and can be answered on a 7-step scale, from "never" to "always."

7.2. Procedure

Data were collected using a quantitative survey within a cross-cultural project. The online study took place on the 9th to 19th July, 2013.[1] Participants were required to have a job with at least 10 working hours per week and have colleagues at work. The workers categorized themselves with respect to their position in the company as leaders or employees.

7.3. Data collecting and participants

All-in-all, 292 respondents from Slovenia answered the questions about HPLC. The respondents chose an appropriate category, i.e., a description of the frequency of the HPLC in their working day from the viewpoint of leaders or employees. The leaders answered how often they pay attention to HPLC in their working day, and the employees answered how often their leaders do so in their working day. Although there were some missing values in the database, the number of responses to each question was sufficient to apply the data analysis methods listed in the next section.

---

[1] Data were collected within the project "Culture4leadership," funded by the Province of Styria, Austria, within the framework of "Grenz-frei" (i.e. "no border") project.

Out of the 292 respondents who answered the questions about HPLC, 74 (25.3%) were classified as leaders and 218 (74.7%) as employees.

7.4. Statistics

The collected data were processed with the Statistical Package for Social Sciences (SPSS; Version 21). As the variables are ordinal, the median was used to measure the central tendency. We also displayed the frequency tables. As the collected data were measured on an ordinal scale, the independent samples Mann-Whitney U test that is the nonparametric equivalent to the parametric independent samples t-test (for example, see [51, 52])was used to verify the null hypothesis: The distribution of the workers' HPLC is the same across categories of position in the company.

8. Results

Tables 2–4 show that the highest middle[2] value of the frequency of the workers' HPLC in their working day falls in the category 5—more often and the lowest middle value falls in the category 3—sometimes. The statements with achieved median 5—more often in both independent samples are listed in **Table 2**.

Statement	**Leaders**		**Employees**		**Asymptotic significance of the difference between leaders and employees**
I answer: *As a leader:* **I as a leader take care that** *As an employee:* **My leader takes care that**	**N**	**Median**	**N**	**Median**	
...it is possible to act in an autonomous and independent way	74	5.00	218	5.00	0.968
...effort isn't unnoticed	74	5.00	217	5.00	0.341
...feedback is received and can be claimed	74	5.00	216	5.00	0.931
...one can rely on that everybody does their work	74	5.00	217	5.00	0.050
...every input is valued	74	5.00	217	5.00	0.907
...quality is valued	74	5.00	215	5.00	0.940
...employees experience in a conversation that one is interested in what they have to say	74	5.00	216	5.00	0.399
...colleagues in a team support each other	74	5.00	214	5.00	0.925

[2] For a data set with an odd number of respondents, the median is the middle value. For a data set with an even number of respondents, the median is the average of the two middle values. To simplify the description of the obtained results, we use the term "the middle value" when talking about the median.

Statement	Leaders		Employees		Asymptotic significance of the difference between leaders and employees
I answer: *As a leader:* **I as a leader take care that** *As an employee:* **My leader takes care that**	**N**	**Median**	**N**	**Median**	
…all employees are treated fairly	74	5.00	214	5.00	0.701
…ideas and suggestions are seriously considered	74	5.00	214	5.00	0.787
…feedback is given in a way, which allows for learning about it	74	5.00	214	5.00	0.846
…work is appreciated	74	5.00	213	5.00	0.930
…feedback is considered	74	5.00	214	5.00	0.868
…employees experience in a conversation that they can speak openly	74	5.00	211	5.00	0.722
…the daily work is consistent with the aims of the company	74	5.00	214	5.00	0.475
…all colleagues of a team talk to each other in an open way	74	5.00	214	5.00	0.697
…there is the possibility to say something when facing uncomfortable decisions	74	5.00	214	5.00	0.757
…knowledge is determining for career and not the relationships somebody has	74	5.00	214	5.00	0.741

N—number of respondents.

Measured on a 7-step ordinal scale: 1—never; 2—seldom; 3—sometimes; 4—often; 5—more often; 6—very often; 7—always

Table 2. The medians of the frequency of the workers' HRLD in their working day—part 1.

Table 3 shows that this middle value (5—more often) was achieved only by leaders, also for the statements that they pay attention that: "Employees are motivated to care for their health," "Nobody must work against their personal values," "One can identify with the values of the organization" and "Health of the employees has a high value." From the viewpoint of employees that their leaders pay attention to these HPLC in their working day, the middle value of the responses to these statements was 4—often.

Employees achieved the median 5—more often only for the statements that their leaders pay attention that: "Criticism is given with appraisal," "It is possible to discuss one's own acting on and off," "Everybody has the necessary latitude" and "In a conversation, employees experience that they are accepted like they are." From the viewpoint of leaders that they pay attention to the first and the third above-mentioned HPLC, the middle value of the responses to these statements was 4—often, and for the second and the fourth above-mentioned HPLC, the middle value was 4.5—from often to more often.

Statement	Leaders		Employees		Asymptotic significance of the difference between leaders and employees
I answer: *As a leader:* **I as a leader take care that** *As an employee:* **My leader takes care that**	**N**	**Median**	**N**	**Median**	
…employees are motivated to care for their health	74	5.00	216	4.00	0.240
…criticism is given with appraisal	74	4.00	214	5.00	0.415
…nobody has to work against their personal values	74	5.00	214	4.00	0.180
…it is possible to discuss one's own acting on and off	74	4.50	214	5.00	0.537
…everybody has the necessary latitude	74	4.00	214	5.00	0.699
…one can identify with the values of the organization	74	5.00	214	4.00	0.272
…health of the employees has a high value	74	5.00	214	4.00	0.526
…employees experience in a conversation that they are accepted like they are	74	4.50	214	5.00	0.933

N—number of respondents.

Measured on a 7-step ordinal scale: 1—never; 2—seldom; 3—sometimes; 4—often; 5—more often; 6—very often; 7—always.

Table 3. The medians of the frequency of the workers' HRLD in their working day—part 2.

Table 4 shows that the lowest median of the frequency of the leaders' HPLC in their working day falls in the category 3—sometimes. It was detected for the statement that they pay attention that "Every mean is shared fairly." From the viewpoint of employees that their leaders pay attention to this HPLC in their working day, the middle value of the responses to this statement was 4—often. For other HPLC items, the median 4—often was achieved in both independent samples.

Statement	Leaders		Employees		Asymptotic significance of the difference between leaders and employees
I answer: *As a leader:* **I as a leader take care that** *As an employee:* **My leader takes care that**	**N**	**Median**	**N**	**Median**	
…interventions against stress are considered	74	4.00	217	4.00	0.439
…the personal career aims have to be consistent with the aims of the company	74	4.00	216	4.00	0.977
…everybody can decide themselves how their work is done	74	4.00	216	4.00	0.923
…there is enough time to finish everything important	74	4.00	214	4.00	0.070

Statement	**Leaders**		**Employees**		**Asymptotic significance of the difference between leaders and employees**
I answer: *As a leader:* **I as a leader take care that** *As an employee:* **My leader takes care that**	**N**	**Median**	**N**	**Median**	
…the working environment is arranged in a health beneficial way	74	4.00	214	4.00	0.523
…there is enough time to finish work	74	4.00	214	4.00	0.200
…performance and success determine career possibilities	74	4.00	214	4.00	0.927
…health of all employees is supported	74	4.00	214	4.00	0.676
…all colleagues of a team can speak openly to each other	74	4.00	214	4.00	0.742
…there is no need for working under high pressure over a longer time	74	4.00	214	4.00	0.481
…every mean is shared fairly	74	3.00	214	4.00	0.195
…employees experience in a conversation that nobody judges them	74	4.00	214	4.00	0.897
…there is closeness between the colleagues	74	4.00	214	4.00	0.276
…the working life doesn't influence the private life in a negative way	73	4.00	214	4.00	0.988
…all colleagues of a team can work together well	74	4.00	214	4.00	0.447
…it is possible to influence resources and space at work	74	4.00	214	4.00	0.420
…there is a limit of work so that someone can follow personal interests	74	4.00	214	4.00	0.984
…everybody uses feedback correctly	74	4.00	214	4.00	0.903

N—number of respondents.

Measured on a 7-step ordinal scale: 1—never; 2—seldom; 3—sometimes; 4—often; 5—more often; 6—very often; 7—always.

Table 4. The medians of the frequency of the workers' HRLD in their working day—part 3.

The results of the two independent samples Mann-Whitney U test written in **Tables 2–4** show that there are no statistically significant differences, regarding the distributions of the workers' HPLC, between the two independent samples of workers: leaders (that they pay attention to HPLC) and employees (that their leaders pay attention to HPLC).

The distribution of the workers' HPLC in their working day is not statistically different across categories of position in the company; **Table 5** presents the relative frequencies only for the statements with different medians in both independent samples (**Tables 3** and **4**). Results in **Table 5** clarify the results of the two independent samples Mann-Whitney U test (**Tables 3** and **4**) even for these statements.

		Frequency—valid percent						
Statement ***I answer:*** ***As a leader:* I as a leader take care that...** ***As an employee:* My leader takes care that...**	**Category**	**Never**	**Seldom**	**Some-times**	**Often**	**More often**	**Very often**	**Always**
...employees are motivated to care for their health	Leaders	5.4	13.5	14.9	14.9	17.6	16.2	17.6
	Employees	9.7	14.8	13.0	17.1	19.0	12.0	14.4
...criticism is given with appraisal	Leaders	1.4	12.2	13.5	24.3	23.0	18.9	6.8
	Employees	5.1	10.7	15.0	15.4	19.2	17.3	17.3
...nobody has to work against their personal values	Leaders	4.1	8.1	14.9	18.9	20.3	25.7	8.1
	Employees	6.1	9.3	20.1	19.6	20.6	11.7	12.6
...it is possible to discuss the own acting on and off	Leaders	2.7	6.8	17.6	23.0	23.0	13.5	13.5
	Employees	4.2	9.8	14.0	15.4	22.0	18.7	15.9
...everybody has the necessary latitude	Leaders	1.4	8.1	21.6	20.3	21.6	18.9	8.1
	Employees	4.2	7.9	16.4	21.0	20.1	19.2	11.2
...every mean is shared fairly	Leaders	10.8	17.6	23.0	12.2	18.9	13.5	4.1
	Employees	10.3	16.8	13.1	20.1	14.0	11.7	14.0
...one can identify with the values of the organization	Leaders	2.7	12.2	13.5	14.9	29.7	20.3	6.8
	Employees	6.1	12.1	15.9	18.2	24.3	15.0	8.4
...health of the employees has a high value	Leaders	2.7	16.2	16.2	10.8	16.2	24.3	13.5
	Employees	8.4	9.8	13.6	20.6	18.2	15.4	14.0
...employees experience in a conversation that they are accepted like they are	Leaders	2.7	10.8	16.2	20.3	21.6	18.9	9.5
	Employees	5.6	11.7	15.0	17.3	17.8	19.6	13.1

Table 5. Frequency table for selected statements about the workers' HPLC in their working day for two independent samples: leaders and employees.

9. Discussion

The presented research results let us confirm both above developed hypotheses. In view of median 5 at the claims in **Table 2**, we can explain that the leaders allow employees their autonomy at work and establish basics for mutuality/inter-dependence. We can also conclude that leaders recognize the employees' achievements and evaluate them appropriately. They also communicate well with employees and give them real-time feedback.

Hence, we can confirm hypothesis "H1: Leaders and employees established positive (social) relations with others." The results presented in **Table 3** show that communication between leaders and employees is appropriate. Employees are more pleased with the communication and feedback than leaders. Namely, not only top managers, but also middle managers were included in the research; the middle and top management do not communicate much. Due to the new legislation [1], WHP is very timely in Slovenia. Leaders are responsible for design and implementation of WHP, but employees should participate, too. WHP requires team work. Therefore, WHP causes improved interpersonal communication and consequently mutual relations. It also supports the systematic approach to stress management, mitie management, career development, etc.

However, the results of the two independent samples Mann-Whitney U test presented in **Tables 2–4** show that there are no statistically significant differences, regarding the distributions of the workers' HPLC, between the two independent samples of workers: leaders (that they pay attention to HPLC) and employees (that their leaders pay attention to HPLC). The hypothesis "H2: There are no statistically significant differences between leaders' and employees' health-promoting leadership culture in their working environment," is thus confirmed.

Namely, according to the legislation [1], one must assess risk at every working place in company/organization, which can negatively impact employees. To minimize these risks, one must implement systematic activities, which positively impact reducing stress and stressors. All activities are performed in the frame of WHP.

Society and companies are dependent on well-trained, highly qualified and motivated employees nowadays more than ever before. At the same time, the potential of workers and their ability to develop became the subject of a new understanding of health, which encompasses both physical and mental well-being, the quality of life and learning.

Companies benefit from WHP: working in better work-environment improves their employees' health and motivation. This consequently results in diminished sickness-related and other costs, better products and services, more innovation and productivity. WHP includes prestige, helping to improve the public image of the given company and its attractiveness for employees, suppliers and customers.

Social insurance organizations benefit, too: successes in health and safety at the workplace diminish costs for the social security (health, pension and accident insurance funds). Healthy working conditions improve the health of the entire population. Less people using medical and rehabilitative services reduce costs in public health service. Companies diminish costs for supplementary wages for sick employees (e.g., continuation of wage payment during sickness), which reduce non-wage labor costs and the contributions to the statutory health insurance funds.

While proposals for prolonging working life are under discussion, in reality more employees leave work earlier for health reasons. The aging of the workforce caused by the demographic change belongs to the major challenges facing the future world of work, which WHP can help to master by helping workers to remain employed throughout their working life.

There are many benefits for employees too: less stress and strain factors and improved well-being and attitude to work. A company is only as healthy and efficient as its employees, hence WHP causes that there are only winners and no losers.

10. Conclusion

This article focuses on the current research state concerning the role of health and safety at workplace and especially on the attitudes about health promotion and leaders' behavior. The literature review supports the conclusion that leaders must be consistent and proactive about incorporating ethics into their leadership agenda, to match principles of social responsibility, including WHP. To successfully implement WHP, leaders must lead their employees in a way, which supports their health. Furthermore, this article reports on the research of the Slovenian leaders' leadership style concerning their employees' health following the new legislation on health and safety at work [1].

10.1. Contributions to theory

The abovementioned empirical research shows that leaders and employees have positive (social) relations with others. The study was performed two years after application of the new legislation for safety and health in Slovenia. We detected no statistically significant differences between two independent samples of workers: leaders (that they pay attention to HPLC) and employees (that their leaders pay attention to HPLC).

For HPLC of leaders, it matters that criticism is appraised, every opinion is shared fairly, one can identify with the values/VCEN of the organization, everybody has the necessary latitude, nobody must work against their personal values, and in conversations employees experience acceptance like they are. These elements can lead to developing new leaders' styles, which can actively promote employee's health. This style must be requisitely holistic and based on real skills and knowledge of leaders and can improve the organizational climate.

10.2. Contributions to practice

Our research enables development of special trainings for leaders, including more holism, humane orientation and health awareness by topics such as communication, organizational culture and climate, motivation, commitment… They could belong to work health promotion.

10.3. Limitations and further research

Further research will be oriented towards the improvement of the measuring instrument for HPLC, convenient for the structural equations modelling in order to make a selection of constructs and indicators that best describe HPLC and to study the relationships between them. Moreover, further research must cover longitudinal researching of HPLC of Slovenian leaders and can enable further development of HPLC. Researches can include measurement of health awareness of leaders and their personal holism. Increasingly important is also leaders' self-assessment.

The obtained research results, however, cover Slovenia with good examples for other countries.

Author details

Simona Šarotar Žižek*, Matjaž Mulej and Vesna Čančer

*Address all correspondence to: simona.sarotar-zizek@um.si

Faculty of Economics and Business, University of Maribor, Maribor, Slovenia

References

[1] Law on Safety & Health at Work in Slovenia ZVZD-1. Zakon o varnosti in zdravju pri delu (ZVZD-1). 2011. Available from: http://www.uradni-list.si/1/content?id=103969 [Accessed 01. 09. 2014]

[2] Wilkinson W,Marmot M. Social determinants of health: the solid facts. 2003. Second edition. WHO Regional office for Europe, Copenhagen

[3] Grawitch M J, Gottschalk M, Munz D C. The path to a healthy workplace: A critical review linking healthy workplace practices, employee well-being, and organizational improvements. 2006. Consulting Psychology Journal: Practice and Research 58(3): 129–147

[4] EU-OSHA. Workplace Health Promotion for Employees. 2010. European Agency for Safety and Health at Work, Bilbao

[5] World Health Organization. 2010. Available from: http://www.who.int/occupational_health/publications/healthy_workplaces_background_documentdfinal.pdf [Accessed 11.10.2014]

[6] Gabrijelčič Blenkuš M, Koprivnikar H, Drev A, Vračko P, Pirnat N, Hočevar T, Vrdelja M, Jeriček Klaščck H, Pucelj V, Kofol-Birc T, Martinovič B, Krajnc I, Martinovič A. Vsevladni pristop za zdravje in blaginjo prebivalcev in zmanjševanje neenakosti v zdravju (Holistic government approach to health and well-being of the population and the reduction of health inequalities). 2012. Institute of Public Health of the Republic of Slovenia, Ljubljana

[7] Šarotar Žižek S, Mulej M. Creating a healthy company by occupational health promotion as a part of social responsibility. 2015. Kybernetes 4(2): 223–243

[8] O'Donnell M P. Definition of health promotion 2.0: embracing passion, enhancing motivation, recognizing dynamic balance, and creating opportunities. 2009. American Journal of Health Promotion 24(1), pp. iv–iv

[9] EU-OSHA. The business benefits of good occupational safety and health. 2008. European Agency for Safety and Health at Work, Bilbao.

[10] Kreis J, Bödeker W. Health-related and economic benefits of workplace health promotion and prevention. 2004. Summary of the scientific evidence. Essen: Publisher: BKK Bundesverband. Available from: http://www.iga-info.de/fileadmin/Veroeffentlichungen/iga-Reporte_Projektberichte/iga-Report_3e_benefits_workplace_prevention.pdf [Accessed 13.09.2014]

[11] Bilban M. Medicina dela za študente tehničDne varnosti (Occupational medicine for the students of technical safety). 2005. Institute of Occupational Safety, Ljubljana

[12] Mulej M. Holistični model dobrega počutja, temelječ in na inovativnosti in družbeni odgovornosti (Holistic model of well-being, based on innovation and social responsibility). 2013. In Premagovanje stresa kot sredstvo za zagotavljanje dobrega počutja (Overcoming stress as a means of ensuring well-being): Scientific mongraphs of project Chance4Change. (193-215). Faculty of Economics and Business, Maribor

[13] ENWHP. Work Health Promotion. 2007. Online. http://www.enwhp.org/workplace-health-promotion.html/. [Accessed 01.07.2015].

[14] Goetzel R Z, Ozminkowski R J. The health and cost benefits of work site health-promotion programs. 2008. Annual Review of Public Health 29: 303–323

[15] Nöhammer E, Schuster C, Stummer H. Employee perceived effects of workplace health promotion. 2013. International Journal of Workplace Health Management 6(1): 38–53

[16] Council Directive EU 89/391/EGS. 1989. Available from: http://eur-ex.europa.eu/LexUriServ/LexUriServ.do?uri=CELEX:31989L0391:sl:HTML [Accessed 08.06.2013]

[17] Chu C, Breucker G, Harris N. Health-promoting workplace-international settings development. 2000. Health Promotion International 15(2): 155–167

[18] Breucker G, Anderson R, Kuhn K. Conclusions. In Federal Institute for Occupational Safety and Health (ed.), Quality Management in Workplace Health Promotion. 1997. Workshop 27–29, November 1996, Berlin/Erkner, Conference Report Tb 81, (pp. 131–144). Wirtschaftsverlag NW, Bremerhaven

[19] Bradbrook S, Duckworth M, Ellwood P, Miedyinski M, Reznolds J. Green jobs and occupational safety and health: foresight on new and emerging risk associated with

new technologies by 2020. 2013. Publications Office of the European Union, Luxembourg

[20] Kalčič M, Lozar A. Zakon o varnosti in zdravju pri delu (Law on Safety and Health at Work) (ZVZD-1). 2011. GV založba; Institute of Occupational Health, Ljubljana

[21] Cergol Lipnik M. (forthcoming): Inoviranje navad za več celovitosti in za bolj varno delovno okolje v lesni industriji po modelu družEbene odgovornosti (Innovation habits for more holistic and more secure working environment in the timber industry on the model of social responsibility). University of Primorska, Faculty of Management, Koper

[22] ENWHP. Toolbox – Details. Health Circles at the Workplace/Gesundheitszirkel im Betrieb. 2014. Online. http://www.enwhp.org/toolbox/details.php?id=29. [Accessed 10.07.2014].

[23] ISO - International Standards Organization. ISO 26000:2010 ISO. 2010. Available from: http://www.iso.org/iso/social_responsibility [Accessed 10.05.2011]

[24] EU. Communication from the Commission to the European Parliament, The Council, The European Economic and Social Committee and The Committee of the Regions. 2011. A renewed EU strategy 2011-14 for Corporate Social Responsibility. COM (2011) 681 final. Brussels, 25.10.2011

[25] Mulej M, Dyck R, editors. Social responsibility beyond Neoliberalism and Charity. 2014. 4 volumes. Bentham Sciences, Shirjah, UAE

[26] Esposito M. Put your corporate social responsibility act together! 2009. Tate Publishing & Enterprises, Mustang, Oklahoma

[27] EU. Commission of the European Communities, 2001: Green Paper on Promoting a European Framework for Corporate Social Responsibility. 2001. COM (2001) 366 final, Brussels: http://eur-lex.europa.eu/LexUriServ/site/en/com/2001/com2001_0366en01.pdf. [Accessed 10.07.2014].

[28] McWilliams A, Siegel D S. Corporate social responsibility: A theory of the firm perspective. 2001. Academy of Management Review 26: 117–127

[29] Waldman D A, Siegel D S, Javidan M. Components of CEO transformational leadership and corporate social responsibility. 2006. Journal of Management Studies 43: 1703–1725

[30] Campbell J. L. Why would corporations behave in socially responsible ways? An institutional theory of corporate social responsibility. 2007. Academy of Management Review 32(3): 946–967

[31] Dankova P, Valeva M, Štrukelj T. A comparative analysis of international corporate social responsibility standards as enterprise policy/governance innovation guidelines. 2015. Systems Research and Behavioral Science 32(2): 152–159

[32] Šarotar Žižek S, Štrukelj T, Mulej M. Enterprise policy for social responsibility and enterprise stakeholder well-being. 2011. In IDIMT-2011: interdisciplinarity in complex

systems. (pp. 233–242). Johannes Kepler Universität, Institut für Systems Engineering & Automation, Linz

[33] Singer P. Practical ethics. 2011. Third Edition. Cambridge University Press, Cambridge, New York

[34] Milner K, Greyling M, Goetzel R, Da Silva R, Kolbe-Alexander T, Pate D, Nossel C, Beckowski M. The relationship between leadership support, workplace health promotion and employee wellbeing in South Africa. 2015. Health Promotion International 30(3): 514-522.

[35] Ciulla J B, Uhl-Bien M, Werhane P H. Leadership ethics. 2013. Sage, Los Angeles, California; London

[36] Luthand F, Avolio B J. Authentic leadership development. 2003. In K. S. Cameron, J. E. Dutton, & R. E. Quinn (Eds.), Positive organizational scholarship (pp. 241–261). Berrett-Koehler, San Francisco.

[37] Maak T, Pless N M. Responsible leadership in a stakeholder society—a relational perspective. 2006. Journal of Business Ethics 66(1): 99–115

[38] Reave L. Spiritual values and practices related to leadership effectiveness. 2005. The Leadership Quarterly 16: 655–687

[39] Fry L W. Toward a theory of spiritual leadership. 2003. The Leadership Quarterly 14: 693–727

[40] Brown M E, Treviño L K, Hartman L P. Ethical leadership: A social learning perspective for construct development and testing. 2005. Organizational Behavior and Human Decision Process 97: 117–134.

[41] Brown M E, Treviño, L K. Ethical leadership: A review and future directions. 2006. The Leadership Quarterly 17: 595–616

[42] Treviño L K, Hartman L P, Brown M E. Moral person and moral manager: how executives develop a reputation for ethical leadership. 2000. California Management Review 42(4): 128–142

[43] Jiménez P, Dunkl A. Healthy leadership in organizations—Introduction of a new seminar concept. 2013. 1st Chance4Change International conference: Overcoming and preventing stress in the individual & Quality of life/well-being in urban areas. Conference Proceeding. Maribor, Slovenia

[44] Jiménez P, Dunkl A, Hofer C, Vogrincic C. Leadership styles in Austria and Slovenia—a new concept of healthy leadership [Proceeding]. 2013. Paper Presented at the 2nd Chance4change International Conference, Maribor, Slovenia

[45] Eriksson A. Health-Promoting Leadership: A Study of the Concept and Critical Conditions for Implementation and Evaluation. Doctoral thesis. 2011. Nordic School of Public, Gothenburg

[46] Gurt J, Elke G. Health promoting leadership: the mediating role of an organizational health culture. 2009. Ergonomics and Health Aspects of Work with Computers Lecture Notes in Computer Science 5624: 29–38

[47] Dunkl A, Jiménez P, Šarotar Žižek S, Milfelner B, Wolfgang K K. Similarities and differences of health-promoting leadership and transformational leadership. 2015. Naše gospodarstvo/Our Economy 61(4): 3–13

[48] Leiter M, Maxlach C. Six areas of worklife: A model of the organizational context of burnout. 1999. Journal of Health and Human Services Administration 21(4): 472–489

[49] Ryff C D. Happiness is everything, or is it? Explorations on the meaning of psychological well-being. 1989. Journal of Personality and Social Psychology 57: 1069–1081.

[50] Jiménez P, Winkler B, Dunkl A. Health Relevant Leadership Dimensions. Development of a Questionnaire to measure Healthy Leadership. 2013. Poster presented at the 16th Congress of the European Association of Work and Organizational Psychology, Münster

[51] Curwin J, Slater R. Quantitative Methods for Business Decisions. 2000. International Thomson Business Press, London

[52] DeCoster J. Testing Group Differences using T-test, ANOVA and Nonparametric Measures. 2006. Available from: http://www.stat-help.com/ANOVA%202006-01-11.pdf [Accessed 02.09.2016]

Section 2

Occupational Medicine

Chapter 6

Is Insulin Resistance Work Related?

Marina Oțelea

Additional information is available at the end of the chapter

http://dx.doi.org/10.5772/66333

Abstract

Incidence of insulin resistance continues to grow, becoming a major public health concern worldwide. Besides the classical risk factors (obesity, unhealthy nutrition and lack of exercise), extensive research about some occupational hazards supports their association with insulin resistance and metabolic syndrome. On the one hand, the classical risk factors for insulin resistance are augmented by the changes of the working conditions: the occupational level of physical activity has a tendency to decrease, reducing its contribution to the overall level of physical activity and favouring a sedentary lifestyle and the occupational stress became the second most common work-related health issue, contributing to the increase of the maladaptive habits, such as unhealthy nutrition. On the other hand, some insulin resistance risk factors are primarily occupational hazards: desynchronization of the circadian rhythm and sleep disruption during the night shifts, workplace air pollution (particles, solvents), heavy metals (arsenic, mercury) or persistent organic pollutants exposure. Meantime, workplaces are excellent settings for health-promotion programmes and metabolic risk reduction, if there is managerial commitment and support. Therefore, assessment of the risk, screening and workplace intervention programmes to reduce insulin resistance incidence should be included in the occupational health service provision.

Keywords: insulin resistance, occupational stress, sleep desynchronization, night shifts, sleep duration, indoor air pollution, persistent organic pollutants arsenic, mercury

1. Introduction

Insulin resistance (IR) is a complex pathophysiological state characterized by reduced target cells responsiveness to insulin. The variety of its manifestations correspond to the complexity of insulin actions in specific tissues; the IR state reduces glucose uptake in muscle and other insulin-dependent tissues, increases lipolysis and fatty acids delivery from adipose tissue, increases hepatic glucose output and sodium retention, impairs uric acid metabolism, enhances vascular endothelium vasoconstriction and inflammation—promoting atherosclerosis and induces a systemic low grade metabolic inflammation. Two are the major clinical consequences of the IR: the cardio-vascular disease and diabetes. There are also significant links between IR and other clinical conditions such as non-alcoholic fatty liver disease, polycystic ovary syndrome and colorectal cancer. Cardiovascular diseases are already the leading cause of death in industrialized countries [1] and diabetes prevalence has doubled between 1980 and 2014. Diabetes is estimated to become the seventh leading cause of death in 2030 [2]. Taken together, all medical conditions related to IR are responsible for a high percentage of the total mortality rate. Therefore, prevention and early diagnosis is a major public health topic.

For many years, IR has been considered a complication of obesity but a non-obese metabolic obese profile has been identified in the past few years, as up to 15–30% of the IR subjects are in the normal range value of the BMI. The complex pathogeny of IR has been explained by complementary or synergic interactions between genetic and epigenetic factors, gut microbiota and environmental factors. As the environmental factors are the modifiable ones, the intervention, including workplace interventions, became nowadays a major health prevention direction.

Insulin resistance is difficult to measure in clinical practice; therefore, surrogates are used for its assessment such as basal insulinemia, the homeostasis assessment index (HOMA-IR), glucose ratio to insulin and the glucose tolerance test. Clinical markers for IR are lipid metabolism impairment (increased triglycerides, reduced HDL cholesterol and increased LDL cholesterol), plasma uric acid increase, renal disease markers (albuminuria) and cardiovascular effects (hypertension and cardiac remodelling). As already mentioned, obesity, particularly the visceral deposition of the adipose tissue, is a risk factor. In different configurations, according to different scientific organizations, these factors define the metabolic syndrome, a more suitable clinical diagnostic. Although I fundamentally agree with those considering that metabolic syndrome is an 'artificial syndrome'—in fact is just a cluster of risk factors—and that diagnostic procedures should be developed for an easier clinical diagnostic of the IR syndrome, this article will have to rely on data accumulated under the metabolic syndrome entity, as it is much more frequently used in the medical literature.

Nutrition, lack of physical exercise are not occupational hazards that are traditionally IR risk factors. In recent research, there is more and more evidence accumulated on how particular occupational risks are involved in IR and this makes the subject of increasing interest for occupational physicians.

2. Epidemiological arguments

An impressive number of studies have emphasized that risk factors for cardiovascular and diabetes are more prevalent in low-educated, low-income population [3–8], particularly related to the high incidence of the negative health habits (smoking, drinking, high fat or low fruit diet [8]). Being a member of a low social class during childhood [9] also favoured unhealthy-related behaviours: low level of sport practice, high fat diet, emotional stress and material unfavourable conditions. In different cohorts, significant differences between men and women [6, 7] were found, which could not be fully explained by sexual level of hormones. Marital status [5] in men appeared to have different influences in developing countries compared to the developed ones: a high risk factor in developing countries and a low risk factor in western countries [4]. The high prevalence in developing countries has been explained by cultural habits such as accessibility to eat out or practice outdoor physical activities and less concern about gaining weight. In women, marital status in western countries is associated with an increased risk.

In the early years of socio-economic development, the changing habits represent a risk factor for IR [10, 11]. The theory of epidemiological transition states that, under development conditions, there is a switch between mortality dominated by infectious diseases to mortality related to degenerative and man-made diseases [12]; conditions related to IR fill in this category, reflecting changes in nutrition and transition from physical activity to sedentariness [11]. In these countries, at the beginning of the development era, the high income class has higher risk [5], as it has access to more resources, moves to more sedentary lifestyle and tries to compensate all childhood economical disadvantages and frustrations accumulated.

Even if there are differences according to the socio-economic development period the country is passing by, there are also many similarities. For example, a comparison between Taiwan and US revealed interesting similar inflammation scores (serum IL-6, CRP, fibrinogen, sICAM-1, sE-selectin and sIL-6R), cardiovascular risk factors (systolic and diastolic blood pressure, total cholesterol, HDL cholesterol, triglycerides, glycosylated haemoglobin, BMI, waist-hip ratio and resting pulse) and hypothalamic-pituitary-adrenal axis and sympathetic nervous system function (epinephrine, norepinephrine, DHEAS and urinary cortisol) in relation to the socio-economic status [13].

Occupation and employment status are major determinants of the economical welfare and unemployment has been linked to metabolic risk in both developed and in under development countries [4, 5, 14]. There is difficult to find common grounds for data reported from different countries, as occupations are differently coded and registered: an US countrywide survey found maximum incidence of risk factors in the farm industry (operators, managers and supervisors), in food preparation and food services, motor vehicle operators and among administrative staff (secretaries, stenographers and typists), and the lowest in professional specialties (particularly in writers, artists, entertainers and athletes) [15]. Several studies have classified occupations in manual (blue collars) and non-manual (white collars, office workers) activities and found either increased or a decreased incidence of metabolic syndrome components in manual workers [16, 17]. These differences were most probably related to the confounders not included in these studies, such as nutrition, smoking, alcohol and level of occupational physical activity.

There are many limitations of these epidemiological studies for a direct application in the occupational physician practice, although the prevalence and social risk factors are important. They give a broader perspective for any intervention proposed. The epidemiological transitions reflect the major trends and do not provide specific information by industry. As far as working conditions are concerned in the transitional period, the main negative factors are the unbalanced working hours to leisure activities, the increased number of jobs with sedentary activity and the increased amount of perceived stress.

3. Occupational hazards

3.1. Sedentary lifestyle

The link between inactivity and insulin resistance has been documented for many years and extensive experimental and clinical research have clarified this relationship in detail. Currently, an inactive person is defined by having less than 30 min/day of at least moderate level of activity, either as occupational or as leisure time activity. As the occupational activities become less physical demanding, it would be reasonable to compensate by more physical activities such as sports practice, gardening and brisk walking [18] instead of increasing the number of hours of TV watching or computer interactions, as we see as a trend nowadays [19]. It is estimated that only 20% of currently industrial jobs demand an average level of physical activity, with a reduction of more than 100 kcal/day in energy expenditure since 1960 [18].

Benefits of exercise in metabolic regulation: Glucose is transported into muscle cell by facilitated diffusion through GLUT4 and GLUT1 channels; GLUT4 is the major transporter. In resting conditions, GLUT4 is located intracellularly and its traffic to the membrane is insulin mediated. Inside the muscular cell, glucose is phosphorylated to glucoso-6-phosphate (G-6P), a compound that cannot transfer outside the cell. During a rest, it is stored as glycogen and during exercise it takes the glycolytic pathway. During exercise, the blood flow increases, thus increasing the amount of glucose available; this creates a gradient for glucose to entry the cell. As cells utilize energy for sustaining muscular contraction, the ratio of AMP/ATP increases, activating the AMP-activated protein kinase (AMPK). This kinase is able to enhance the transport of GLUT4 to the membrane in an independent insulin manner. AMPK activity is also increased by leptin, through indirect hypothalamic action and adiponectin, by a direct action. In obesity, resistance to leptin diminishes AMPK activity [20]. Exercise increases GLUT4 transport and fatty acid oxidation, improving insulin sensitivity. Acute changes depend on the energy balance: if this is positive, the carbohydrate oxidation will prevail and if it is negative, the fatty oxidation will be enhanced [20]. In order to reduce fatty mass, a negative balance is also requested. The type of exercise also influences the energy substrate utilization: endurance training increases fatty acid oxidation while resistance training, increases glucose uptake and utilization; most probably, this is due to the increase in muscular mass [21]. Lipoprotein lipase increases after training and maintains a higher level afterwards; the higher the level of exercise, the longer will be the duration of the high level of lipoprotein lipase and the preventive action. Chronic training contributes to increase the number of myofibrils and their oxidative capacity. Training has also a role in restoring mitochondrial content in ageing populations or in different pathological conditions [22–25] and in activating satellite cells [26]. The mitogenesis effect increased with 25% when calorie restriction was associated [27]. In previous sedentary obese men and women, training increased the mitochondrial electron transport and decreased IR.

Exercise partially counteracts other risk factors, such as a fatty diet. High lipid intake increases the fatty acids content in muscular cells, β-oxidation, acetilCoA level, NADH/NAD ratio (slowing of tricarboxylic acid cycle) and leads to citrate accumulation. Citrate inhibits muscle hexokinase and reduces the formation of glucoso-6-phosphate; the glucose traffic through

http://dx.doi.org/10.5772/66333

GLUT4 channels is, therefore, reduced, contributing to chronic hyperglycaemia. This processes favours fatty acid utilization and counteracts the insulin metabolic switch. A vicious pathological cycle is created between decreased sensitivity of GLUT 4 and reduced uptake of glucose. Some benefits of exercise are explained by disrupting this cycle. Training even reversed the other negative risk factors, such as obesity influence or small size at birth [28].

Relationship between sedentary lifestyle and obesity: Sedentary lifestyle infers a low energy expenditure that should be adjusted by a balanced reduction in food intake, but food intake is driven by many other factors compared to expenditure in abundant, accessible nutrient societies. In such context, calorie intake becomes easy and higher than requested for thermoregulation and metabolic purposes. In a strict mathematical formula, this will lead to increase in body weight. High calorie intake is a consequence of high fat and/or high-carbohydrate diets. Randle's hypothesis of fatty acid toxicity [29] is a classical explanation of IR induction by excessive fatty diets. High circulating level of fatty acids inactivates insulin cellular signal and impairs the switching of the metabolism from fatty acids to glucose, with progressive accumulation of intra-cytoplasmic lipids in insulin sensitive cells (including β-cells); the lipid accumulation, mainly of ceramides, induces apoptosis. In muscular cells, the increased level of fatty acids inhibits phosphorylation of insulin receptor substrate 1 and activation of phosphoinositol 3 kinase, with a decrease in glycogen storage. High carbohydrate diets, particularly of those with high glycaemic index, desensitise glucose receptors and increase oxidative stress on β-cells that ultimately lead to insulin depletion. High fructose intake feeds the gluconeogenesis and the lipogenic pathways [30] in hepatic cells and by depleting the cellular ATP content and gives to the brain a signal of energy deprivation, favouring food seeking behaviour. Increased hepatic glucose output maintains a hyper-insulinemic state with higher fat storage in adipose tissue.

Sedentary life decreases the proportion between the fat mass and the free fat mass. According to the energy balance concept, it would be expected, as the total body mass increases, that the total energy consumption increases, from the basal metabolism increase. The energy expenditure is dominated (with exception for highly competitive professional athletes) by the resting metabolic rate and the variation in size of the free fat mass that explains 60–80% of the variation in basal metabolism [31], with an approximated 5:1 ratio between free fat mass and fat mass [32]. Energy balance is a remarkable concept, but cannot fully explain the complexity of the metabolism regulation and of the metabolic flexibility, governed by insulin. In sedentary adult with positive energy balance, the proportion of fatty mass increases gradually; even if the adjusted increase of the basal metabolic rate would be proportionate with the total mass, this increase will not compensate the level of calories intake, because fat mass has lower basal metabolism than skeletal muscle and fat will continue to accumulate. Obesity, as defined by a quantitative increase of fat tissue or as proportion between fat mass and fat free mass [33], is a direct consequence. The abnormal proportion and not the BMI is, in fact, related to the metabolic risk factors. This explains, at least partially, the cardio-metabolic risk of the non-obese metabolically obese population. This increase in quantity is also associated with qualitative impairments, a phenomenon described as adiposopathy or 'sick fat' [34]. The increased level of leptin and of other inflammatory cytokines, the decreased level of adiponectin, switch of the macrophages to the M1 pattern and to a pro-inflammatory status and lower capacity of oxidizing fat are the main characteristics of the metabolic pattern of adiposopathy. Muscle is the main consumer of glucose and fatty acids and by that, the major contributor to the lowering of glucose and of the fatty acid plasma levels; if free fat mass (mainly muscle mass) is reduced due to the sedentary lifestyle, the risk of adiposopathy increases.

Occupational level of physical activity has a general tendency to decrease. This reduction in contribution to the overall level of physical activity explains that a significant IR risk was associated with leisure time sedentary behaviour but not with occupational activity in general population [35]. However, when looking more closely, it becomes obvious that several professions are at higher risk: for example, VDU operators in call centres [36] or bus drivers [37] had a higher OR for metabolic syndrome, after confounders such as smoking or leisure time activity were excluded. Comparison of sedentary lifestyle has been highly suggestive between different professions and even among same profession that has specificities according to national regulations: for example, police officers and administrative clerks have the higher risk than fire fighters in states where training programmes are strictly supervised but not in voluntary forces (US) [38].

During work, the sedentary activity can be more sustained (more than 30 min) than during non-working hours and this has also an impact [39]. Beneficial effects in adiposity measures, triglycerides and 2-h plasma glucose were obtained even in sedentary persons with more interruptions in sedentary time [40].

Sitting for more than 1 h reduces the lipoprotein lipase activity [41] and the non-exercise activity thermogenesis [42]. Adjustable workstations might be a solution for prolonged sitting activity. In a recent systematic review, the majority of prospective studies found that occupational sitting was associated with a higher risk of diabetes and mortality and to some degree of cardiovascular risk. These apparent conflicting results between the two consequences of the IR could be related to the heterogeneity of study designs and to the inconsistency and sufficiency of data recorded on overall physical activity. Possible other confounders are diet, tobacco, alcohol and high energy food consumption, as sitting for long hours increases the chance of negative habits to develop. When the total physical activity was considered, the association between sitting and metabolic risk lost the statistical significance [43].

Clinical approach: Clinical evaluation of the physical activity is rather easy and should be part of the routine examination. BMI is also very frequently recorded and, even if it has the above-mentioned limitations, it has an orientation value. In order to be more specific about the metabolic risk, abdominal circumference could be added to the physical examination, particularly to the non-obese persons, defined by the BMI index. More accurate data about fat mass proportion are obtained by bioelectrical impedance analysis, but they are not used in routine clinical examinations.

As high physical activity has numerous benefits, it should be encouraged to everyone, according to its age, health status and preferences. There is enough evidence that even small increases, if they are sustained and become part of the routine, have high medium-term impact as primary or secondary preventive measures. If sitting is the main problem, an ergonomic intervention could be beneficial.

http://dx.doi.org/10.5772/66333

3.2. Sleep and night shifts

Sleep duration and sleep architecture are important indicators of individual health. When normal sleep is impaired, alertness, memory, learning, thermoregulation, respiratory, metabolic and cardio-vascular regulation are also disturbed. This broad spectrum of effects is one of the consequences of the complex connections between the hypothalamic clock and different nuclei in the central nervous system [44] and of the interaction between the central and the peripheral circadian regulators. Assessing sleep duration, sleep-wake cycle and sleep disorders in the occupational medical practice is very important, because they influence significantly the fitness for work particularly in high risk jobs for accidents (professional driving, traffic controllers, working at height, etc.).

Normal sleep regulation: Sleep onset and duration are the core elements of sleep-wake regulation. Sleep-wake regulation is currently explained by two biological models: the sleep homeostasis and the circadian clock. Sleep homeostasis refers primarily to the variation of sleep pressure, while the circadian clock concerns the activity of the hypothalamic pacemaker, acting independently of the prior sleep period [45]. Although interrelated, their regulation pathways are different: sleep pressure is mainly dependent on the existing sleep debt and is the physiological contributor to the recuperation of this debt. If during the previous day the sleep duration was shorter, the sleep homeostasis will intervene to correct the deficit.

The pacemaker of the clock automatism is genetically set to a circadian rhythm and is controlled by the environmental factors. The circadian period (the interval that covers a full cycle of sleep-wake, not influenced by external constraints) is the *tau*; *tau* varies in healthy individuals between 24 and 25 h and reflects the automatism of the suprachiasmatic nuclei processes (is circadian clock dependent). In healthy adult individuals, periodicity of sleep duration waxes and wanes ranges broadly (around 2–18 days), indicating that for a specific individual the number of hours slept varies in consecutive days, having a repetitive pattern of duration. For example, for the 5 days pattern, the number of hours slept in day 1 is the same as the one in day 6; between day 1 and day 6 there is an increase and then a decrease in duration of sleep. This inter-individual variation is determined by the different build up sleep pressure in healthy individuals and is dependent primarily to the sleep homeostasis. This variance will impact on different ways the adaptation to night shifts and to longer working hours of different individuals.

Under a certain extent, the clock is self-regulated by its genomic functionality. The activation of the hypothalamic neurons starts with the transcription of the *Clock* and *Bmal 1* genes that have circadian variations, related to day-night transitions; their products form a heterodimer (CLOCK-BMAL1) that binds to the promotion site of *Period 2* (*Per2*), *Cryptochrome* (*Cry*). The cytoplasmic levels of PER2 and CRY regulate the transcription induced by CLOCK-BMAL1 heterodimer through a negative feedback loop. Besides this auto-regulated feedback loop, *Clock* and *Bmal1* are also influenced by a secondary control loop, acting on two nuclear receptors, the *Retinoid-related orphan receptor* (*Ror*) and the *Rev-Erb receptor*. While Ror is a transcription activator, *Rev-Erb* is an inhibitor one [46]. PER and CRY reach a maximum at the normal transition from day to night maintaining the negative feedback loop of the clock and contributing to the sleep onset. BMAL1 reaches the maximum at the transition from day to night, at waking time. In order to generate the level of BMAL1 requested for waking, ROR and REV ERB maxima occur between that of BMAL1 and PER-CRY.

Dark is the physiological stimulus of melatonin, although melatonin increases during the evening even in blind persons [47]. While its major role is to induce sleep, melatonin has also other functions such as heat loss or glutathione synthesis inducer [48]. It also influences the immune response and the glucose metabolism.

Sleep cycles have normal variation during the night: in the first part of the night, there are more non-rapid eye movement (NREM) periods than in the last part; stage 1 of sleep diminishes (in percentage time of the total cycle) in the second part of the night and stage 1 is no more present after the first 4 h of night sleep. During nightshifts or sleep interruptions, the normal sleep architecture is modified and explains the tiredness people feel, after a working night, even after an equal number of hours of recovery sleep.

Normal sleep duration: A large European study revealed significant differences between countries, related mainly to cultural habits than to environmental ones (sunlight or temperature variation) as winter and summer sleep duration showed no significant variation by country [49]. On average, adults sleep 7–7.5 h/night [50]. In adults, adjusted prevalence of short sleepers (<5–6 h/night) or long sleepers (>10 h/night) is 0.52 and 0.64%, respectively [51]. Duration of sleep is influenced by genetic, personality traits and social habits. Long sleepers had higher neuroticism scores and lower education levels than short sleepers [51, 52]. Short sleepers are more frequent in adults that have not sleep alone in their childhood, as toddlers or pre-schoolers [52]. Familial nocturnal short sleepers have been identified in a mutation on the gene *DEC2*: in persons caring this mutation, the whole duration of sleep in 24 h is less than 6 h [53]. Shorter sleeping time was associated with the total time spent sitting and with increased screen time in a large cross-sectional European survey [54], linking this risk factor with sedentary lifestyle and with a long time of computer interaction.

Metabolic effects related to sleep regulation: Hypothalamic clock does not regulate just the sleep-wake status; it is the central driver for the peripheral clocks correspondents in colon, stomach, pancreas, liver, adipose tissue and muscles: through this integration, certain metabolic pathways have circadian fluctuation. It is estimated that 10–15% of the gene transcripts have rhythmically 24-h cycles; besides the genes expression, translational changes have also circadian fluctuations. There is a normal variation of the peak transcripts during day (e.g. proteins involved in inflammation and hormonal response) or night (e.g. proteins involved in ribosomal and epigenetic processes such as methyltransferases) [55]. In twin-based genetic studies, heritability is found responsible for 33–55% of the sleep duration; interestingly, heritability influence increases when twins are living together for a longer period in their life [56].

Besides the central clock influence, there is also an intrinsic circadian variance of the metabolic genes, as response of the Clock/Bmal1 or Per 2 variation of expression. All hormones involved in glucose regulation (cortisol, catecholamine, glucagon and insulin) have circadian peaks. There is significant increase in insulin sensitivity during the early nocturnal sleep and a returning to the pre-sleep values during the late phase of the sleep [57]. In animal studies, melatonin decreases insulin and increases glucagon secretion in pancreatic cells; high insulin levels, as reached in type II diabetes, decrease the melatonin level [58]. In humans, several genetic polymorphisms of the melatonin gene are risk factors for IR. A later rise of the morning melatonin peak is encountered in persons with the GG allele of the melatonin gene and this time corresponds to the usual timing for food intake. The later than normal melatonin raise deregulates glucose metabolism, as it counteracts the nutrient-related signals transmitted to the pancreas cells. Other genetic variants, as the *MTNR1B* gene variant, increase the number of melatonin receptors expressed by β-cells with negative impact on the insulin secretion. Subjects carrying this allele, receiving 4 mg melatonin/day, had, after 3 months of treatment, significantly less insulin secretion and were more IR than controls [59]. Possible future testing of these genomic characteristics will allow occupational physicians to distinguish the persons that will have more difficulties in adapting to working in shifts and those that are at risk of developing metabolism disturbances, including IR.

Negative influences of work schedule on insulin resistance

1. *Short sleep duration.* Short-term sleep loss could be perceived as a metabolic stressor: it increases metabolism but also the anti-oxidant capacity [60]. These adaptations are surpassed in sleep deprivation, when major disturbances in hormonal regulation occur. Glucagon and cortisol are lower after a night of sleep deprivation [61]; there is also a lower response in adrenaline to hypoglycemia, explaining both the hunger and the fatigue syndrome in the day after. The day after sleep deprivation is characterized by both autonomic symptoms (anxiety, sweating, irritability, hunger) and glycopenic symptoms: dizziness, blurred vision and difficulty in thinking. After four nights of sleep deficit, the insulin signal pathway was reduced by 30% in adipocytes [62]. Cortisol levels increase just after one night of sleep deprivation. Even if the glucose tolerance tested by I.V. glucose tolerance test was similar with the reference group, the short sleepers had a reduction by 40% of the insulin sensitivity, both in the first and in the second phases of response [63]. Another strong argument of the influence of sleep on the IR is the reduction of HOMA-IR after an increase of amount of time slept by 45 min/night, a rather cost-effective intervention for prevention of IR [64].

   The work time arrangements are very important for maintaining a physiological recovering sleep. Increased risk for metabolic syndrome was not constantly found in relation to working in shifts. However, alterations in glucose and lipid metabolism, observed in experimental settings [65, 66] of night shifts combined with a total reduction on sleeping hours (less than 5 h/day) and adverse nutritional habits [67] significantly increase the risk.

2. *Desynchronization of the circadian rhythm.* Another consequence of working in shifts is the shifted circadian rhythm (desynchronization, misalignment). The desynchronization increases the effects of sleep deprivation; the insulin sensitivity is twice reduced if restriction is combined with 8.5 sleep delay (circadian misalignment). After 10 days of longer day duration (duration of 28 h) a significant increase of glucose, decrease of leptin, reversed cortisol cycle and an increase of mean arterial pressure [68] have been recorded. A decrease in insulin sensitivity and leptin secretion are also common [69].

   The metabolic effects of the desynchronization on the circadian rhythm have been explained by the conflict between the disturbed circadian variations of the peripheral clocks and the feeding stimulus. In general, peripheral clock shifts at a slower rate than the central clock. The circadian variation of different peripheral factors involved in metabolism regulation such as adipokines [70], pancreatic secretions, PPARγ PGC1α is supported by numerous experiments. Besides this circadian variation, their expression is also triggered by the nutrition intake. For example, leptin plasma levels depend on meal time; irregular time of nutrient intake is a risk factor for the leptin resistance, a common finding in IR [71]. Pancreatic cell possesses self-sustained circadian genes and protein oscillations of the transcription factors clock bmal1 [72]. PPARγ, a regulator of adipocyte differentiation and secretion and of insulin signal transmission, has circadian oscillations [73] but it is also activated by fatty acids. PGC1α coordinates genes in lipid and glucose metabolism, the PAI-1 promoter and the thrombocytes activation [30]. It is stimulated, in heart and liver, by fasting [74]. What are the specific interferences between the intrinsic oscillation of expression of these proteins and the environmental factors (mainly nutrition and light) are currently a subject of intense research.

Social and psychological factors are other important modifiers of sleep homeostasis and even of the pacemaker control, with specific markers at the cellular level. For example, in forced-desynchrony experiments in which the sleep-wake cycle and the associated fasting-feeding and dark-dim light cycles are scheduled at a longer *tau* than 24 h, disrupted transcription of the clock genes expression, epigenetic modifications of chromatin, impairment of the RNA polymerase, modification of the regulation of transcription and translation of heat shock proteins and temperature-sensitive RNA binding proteins (CIRBP and RBM3) have been recorded [55].

The higher weight gain of the night eating syndrome is another consequence of the desynchronization of the circadian rhythm. A plausible biological explanation is that the nutrient-induced stimulus is acting out of the phase of the physiologic oscillations of these proteins. If, for example, melatonin secretion is delayed, there will be a decrease time between wake time and circadian melatonin offset phase that will increase the eating behaviour. The modification of appetite could also be related to changes in ghrelin and leptin levels in short sleepers, significantly increasing BMI [75]. Changes in preference for salty or high carbohydrate foods, irregular meals and snaking behaviours [76] have also been reported.

3. *Sleep disruption* is well studied particularly in sleep apnea, but in this specific sleeping disorder the intermittent hypoxia was the core mechanism for the explanation of the majority of the cardio-metabolic influences.

During the night shift, sleep disruption cannot be avoided. It certainly affects the quality of sleep and the sleep architecture (the proportion between REM and NREM cycles and their duration). Sleep (particularly REM cycles) activates transfer from the short- to long-term memory and allows, from the occupational perspective, to integrate new tasks of the job; therefore, time to recovery is an important factor influencing the stress level and the employees' productivity. Response to meaningful stimuli is different during sleep: for example, a person will wake up to a lower voice stimulus while hearing his own name versus hearing someone else's name [77].

The irregular sleep-wake cycle has been associated, in an ageing population, with higher cardio-vascular risk, independent of other risk factors. The metabolic consequences of frequent sleep disruption are similar to those of sleep deprivation.

Sleep latency may become problematic after frequent disruptions of the sleep and this affects the duration of sleep after the shift, acting in synergy with the short sleep duration.

Clinical evaluation: Anamnesis should focus on the duration, difficulty in sleep initiation and on the napping habits. Complaints such as diminished intellectual performance, tiredness, loss of interest, reduced level of attention or concentration on tasks, increased number of mistakes and work accidents should always raise the awareness about a possible sleep disorder. Dangerous habits, such as smoking, alcohol and auto-medication for sleeping pills might be the clinical manifestation of the sleep problems.

In clinical practice, a *structured interview* is very helpful for all employees at risk. In general, the following items are investigated: bedtime (bedtime resistance, sleep onset delay), sleep behaviour (amount, snoring, leg movements, number of waking episodes) and morning wake up (time, self-induced, sleepiness during the day). The sleepiness level, for example, can be evaluated using different scales. The extreme chronotypes are assessed with the Horne-Östberg morningness-eveningness questionnaire [78]. The major items used in this questionnaire are: time of getting up and of going to bed, level of alertness, the best interval during the day for occupational or recreational activities and the perceived difficulty to perform a certain activity at different hours. *A sleep diary* for a minimum 2 weeks is a very good tool for monitoring sleep duration, day or night distribution, weekly distribution, onset and wake up time. The non-recuperated sleep debt might be the underlying cause of the complaints.

Determination of the specific sleep pattern (periodicity of sleep duration waxes and wanes) would be ideal for pacing the shifts. The *maintenance wakefulness test* is sometimes necessary to assess the ability to stay awake in soporific conditions (quietness, dim light, confortable temperature). It is measured using the polysomnography method under standard conditions [79] and is an objective measure of the inability of some patients to maintain normal alertness during the daytime.

The first step in any sleep disorder without organic substrates is to eliminate contextual factors for the sleep latency or delayed sleep onset. Sometimes, just social adjustments or introduction of the basic recommendations for an efficient sleep routine will be enough (relaxing activities 1/2 h before going to sleep, avoid TV, computer screen light, avoid intense physical activities 3 h before going to bed, avoid caffeine containing beverages in the second part of the day, etc.). In other cases, avoiding shift work become mandatory, at least for a certain period of time, enough for recovery.

A special issue is related to the existing diabetes or cardio-vascular disease that might be aggravated by this working schedule. In some countries, national regulations allow the occupational physician to demand the employer another kind of working time. Whenever this is possible, it seems to be the best option.

3.3. Stress

There are many *definitions of stress*. From the biological perspective, the process underlying stress is initiated by stressors and implies a stress-response (bodily reaction to stress, both psychological and physical) that also creates an emotional impact. If this impact is emotionally positive, the perception is of a positive or motivational stress; the negative emotional impact defines the distress. In response to chronic stressor stimulation, activation of the hypothalamic-pituitary-adrenal axis and of the sympathetic nervous system increases cortisol and adrenaline systemic levels, heart rate and blood pressure. These biological effects have been used to measure, objectively, the level of stress. Chronic high levels of counter regulatory hormones (cortisol and adrenaline) are a plausible biological explanation of the insulin resistance related to stress. Indeed, in individuals with metabolic syndrome, levels of cortisol metabolites and catecholamine are found significantly higher than in controls, with a reduced cardiac frequency variability related to the higher sympathetic tonus, and high levels of inflammatory markers (protein C and IL6); the psychosocial factors are estimated to explain around 13% of these effects [80]. Negative emotions have an acute impact on the inflammatory and vasoconstrictor tonus; an interesting comparison between acute myocardial infarction precipitated by a stressful event and the ones without a distinguished negative event showed a significantly higher monocyte chemoattractant protein-1 (MCP-1) and endothelin-1 (ET-1) in the first group [81]. Distress also influences the spontaneous individual strategies to cope with stress, leading to unhealthy behaviours such as smoking, bad nutrition, low physical activity and addictions [82].

Occupational stress is considered the second most common work-related health problem, affecting 28–40% of the workforce [83, 84]. Several conceptual models have been developed to analyse the level of occupational stress. The theory of Karasek et al. based on the job demand and the control power or autonomy [85] focuses mainly on the personal control and short-term stressors. Siegrist et al. [86] created a model based on the effort -reward balance, with more emphasis on social equity and on long-term perspective [87]. A third model, the job demand-resources model, has been proposed by Bakker and Demerouti [88], challenging previous approaches for treating the diverse working conditions in a similar and general way; this model considers that even if job demands are relatively high, they will not generate distress if the organization provides the functional frame for achieving the goals, reducing by that the physical and psychological cost of the employee. Even if there are certain overlaps in these models, their focuses are different: while the Karasek et al.'s model focuses on the psychobiological impact on the threatened control, Siegrist et al. changed the accent on the threatened reward and Bakker and Demerouti have highlighted the threatened support.

In clinical research and practice, the measurement instrument of the perceived stress is the questionnaire. Depending on the conceptual model used, the items are different: for the first model, the quantity of work, the time constraints, the autonomy in making decisions, including the creativity and the possibility of skills development are the main items. The effort-reward model uses questions to assess the time constraint, the responsibility, competitiveness and fairness of treatment inside the organization, the job promotion perspectives, respect and esteem, job status, security of employment and external motivators (adequate salary). The demand-resource model claims a specific analysis by the type of activity and organization resources and hence, there is no a general questionnaire proposed. These instruments have been enriched, in certain studies, with specific items related to sleep disorders, fatigue, sufficiency of recovery time, and posttraumatic stress (the psychological injury risk indicator) [89], and they continue to evolve, creating much difficulty in the comparison of data.

http://dx.doi.org/10.5772/66333

Occupational stress is closely related to job dissatisfaction. Job satisfaction has a multitude of determinants, from job content, to values and leadership of the organization, management style (participative or authorship) and organizational culture (supportive or punitive, leadership role model, carrier paths definition, co-worker support, salary benefits, recognition, work-life balance, etc.). Nowadays, when competitiveness demands frequent improvements in the organization, proper change management processes became increasingly important in maintenance of the job satisfaction. Change is stressful if there is not enough planning, supportive leadership and if it is very frequent, creating the feeling of instability [90]. It is the management duty to find possibilities for job enrichment, job enlargement or job rotation in the process of change to increase the internal motivation and the meaningfulness of the job. But the rapid development of work and constant training create high pressure on employees and reduce family and recovery/recreational time, no matter how motivated the employee is. It can be only partially compensated, sometimes, by a perceived increase of the quality of family time through the actualization of the new skills in different social contexts.

Employees quit managers, not companies but they also leave if there is a misfit between their competencies and skills and the tasks and goals requested (poor recruitment process), if there is role ambiguity or too little decision power. Work time arrangements have several aspects in which the management of the organization can intervene to reduce stress and to increase job satisfaction and productivity such as shift work scheduling, number and flexibility of the work hours and overtime. By simply forward-rotating the shift system, the mean number of days on which the workers reported sleepiness, the heart rate at rest and the systolic blood pressure decreased significantly [91]. Longer working hours lead to increase in smoking, drinking and weight gain [92], affect the need to recovery time and create psychological distress [93]. The *Karoshi* phenomenon ('death from overwork', in Japanese, associated with more than 60 h/work month) is rarely recognized as such in western societies [94], but the silent, progressive association between insulin resistance and stress might be, interpreted, in fact, as part of the same picture.

Particularly in services, flexibility in both time and location is a new way of working that shifts the organization from standard, repetitive procedures to project-based short-term activities. In certain developed countries, such as Sweden, only 16% of jobs are regulated in terms of time, space, process flow and coordination, performance evaluation and peer interrelation, with differences between private (more flexibility) and public sectors, activity domains (services more than production) and gender: women more frequently employed in positions with higher levels of regulation than men [95]. Flexibility has certain advantages for stress reduction: it fulfils the need for autonomy and decision empowerment and gives the employee the perception of the self-determination of his working time, in terms of duration and daytime distribution. On the other hand, flexibility creates an opportunity for the employee to stay in a permanent 'switch on' work activity, predisposes to isolation, lack of supervision and peer support and reduces the development of the interpersonal skills and relations. These can lead to burnout or to unhealthy behaviours (e.g. snaking, increase consumption of sweet beverages, smoking) [96]. Family working balance is more and more important in dual-career couple societies and effort recovery time and psychological detachment from work are central in maintaining this equilibrium.

Relationship between IR and occupational stress: To answer this question, a comprehensive systematic review [97] of the prospective studies linking stress and metabolic syndrome found a positive relation. Gender seems to be important in terms of stressors: for women, the main risk factor was job strain, whereas for men the feeling of justice at work was the main protective factor. When assessing the individual components of the metabolic syndrome, weight gain was inconstantly associated with job stress, but low justice at work was associated with high triglycerides in men and job strain with high cholesterol in both sexes. Development of diabetes in men was in direct relation with higher demands. Job insecurity and low feeling of justice in men and effort-reward imbalance in women were associated with an increased high blood pressure risk. Low effort and low engagement lead to increased dissatisfaction towards job and in medium term might lead to a certain degree of depression. At the opposite end, the low effort and low engagement are overcommitment and this attitude also amplifies the adverse health effects; overcommitment is expressed by a continuous thinking of work during the normal social/relaxing time, with an impossible switch off from work. From psychological perspective, it underlies the need for approval and recognition at work [98]. The most frequent manifestations of overcommitment are burn out, psychological distress, musculoskeletal complaints and self-rated poor health [99, 100]. In a large Dutch study, overcommitment increased by more than 20 times the risk of emotional exhaustion in workers recording an imbalance between effort and reward compared to those without overcommitment [101].

Direct life-threatening jobs (policemen, firefighters, sailors) are well-known stressing conditions. Policemen has a high score in demand-control-support scale and in the effort-reward imbalance scale and these scores could explain the more than doubled OR of metabolic syndrome, particularly for the hypertriglyceridemia component [102, 103]. Working in confined spaces is another well-recognized stressor: after only 5 months of sailing around the world, in a rather small ship with only 16 persons an equipage, BMI, HbA1C, basal insulinemia and triglyceride level [104] significantly increased. In these particular working conditions, the circadian rhythm modification, the changing working environment, the rapidity in decision process, the loss of control on own private life and private activities and the anxiety in front of unknown future conditions were considered as additional stressors.

But occupational stress does not require only extreme conditions. A large British study, including more than 10,000 civil servants, using the demand control evaluation method in repetitive measurements, concluded that more than 14 years of chronic stress exposure increases the risk of metabolic syndrome by two in men and by five in women [105].

If job might be stressful, nevertheless loosing job or job insecurity is another very stressful event. This association is supported by many studies [106–108]. Moreover, the relationship between the numbers of financial negative events in one's life was independently related to insulin resistance in a large, general population study in Finland [109].

Health risk factors (smoking, drinking, low physical activity, increased BMI) are more frequent in public sector employees perceiving low rewards or low occupational effort [89]. While low-reward perception is intuitively link with dissatisfaction, low effort should be expected to generate the opposite; however, in a large-scale meta-analysis, it has been found that low effort and low engagement reduce the job satisfaction and increase the level of depression [110]. These occupational stressors generate coping behaviours for the job in satisfaction and depression. However, in both prospective and cross-sectional studies, the strength of the association could be minimized by the variety of individual reactions to stress, reducing the statistical significance for each behaviour.

Stress and nutrition relationship: As nutrition is so important in the insulin resistance physiopathology, the effect of stress on eating behaviours has been largely studied. In an ideal homeostatic balance, appetite and eating should be driven by the internal eating cues (hunger and satiety). But food is also a basic primary reward and in our food abundant modern society, numerous other external factors trigger eating. Emotional food choices are certainly influenced by the level of the stress hormones; other factors, such as gender, age and cultural modulates this behaviour. Female tends to be influenced more by the negative emotions. For example, in young women, binge eating has been related to executive function impairment such as cognitive impulsivity, insufficient planning and higher levels of cortisol release in relation to stressors. The tendency towards depression is also high among them [111]. On the contrary, men are increasing their portions, when experiencing positive emotions [112].

Stressed students and adolescents generally eat less healthy [113], and this attitude is also more influenced by negative emotions [33]; body image is the main concern [114] at this age that opposes the emotional food intake. Adults may improve emotion control and, in general, increase the health concerns about food; in this context, in healthy subjects, the positive emotions become more powerful in leading eating behaviour. Apparently, even liking for specific taste changes with age: salt liking increases and sweet taste decreases [115]. Emotional eating is frequently related to sweat taste liking that is more expressed in women [115, 116].

Stress and eating dysregulation or a poor differentiation between internal eating cues (hunger, hypoglycaemia) and external stressors are closely related and may be expressed in different personalities either as overeating or as fasting. Eating dysregulation is more frequent in persons having difficulties of identification own feelings and emotions (alexithymia) [117]. High-caloric foods (snacks) are convenient food for this type of behaviour; indeed, in experimental induced stress, people often consume more sweet fat foods than usually [118]. There is a biological explanation of this consumption, as such foods increase the serotonin level that improves mood. The mood improvement creates a positive, but medium- and long-term unhealthy, feedback loop, that contributes to permanentize the behaviour. Emotional eating has been explained by inadequate parenting practices during childhood: an increase control and restriction of certain foods for fear of weight gain (the 'dietary restraint model') or usage of foods to emotionally comfort ('affective state model') are influencing the adult emotional eating and create a significant role for the food-reward behaviour in adults [119]. Recent studies highlighted the genetic role: an increased number of central dopaminergic receptor 2 and a melatonin receptor allele are a risk factor for emotional eating and energy intake under stressful conditions [120, 121].

Even if link between stress and insulin resistance seems to be well argued by the existing data, some limitations of these studies should be highlighted. The comparability and reproducibility of data are difficult because there are a variety of instruments used to measure stress. Another limitation is that most of these studies (particularly the large ones) are studies from the public sector, and might not be fully transposed into the private sector, where competitiveness might be higher and job employment less predictable. Even less is known about small or family enterprises that have specificities in organization, human interrelation and high interference between family and occupational stress. Personal (non-published) experience with local small-medium size firms showed that where there is tendency of preferred employment of relatives, there are additional stress factors, work inequities and general level of stress perception inside a firm.

Fortunately we still have large prospective studies confirming the risk between different components of the metabolic syndrome and stress perception. Even so, the methodology of these studies could be challenged: in many studies, there is no follow-up of the drop off (quitting the job, for example, because of the high stress, weather there is a job demand/job control induced distress or an effort-reward imbalance). If we would have the methodology to estimate in a proper way the healthy worker effect maybe the risk will be even higher, but this a common bias with all occupational risk hazards. When assessing the nutritional effect of job stress, almost all studies examined the increased in BMI and none have considered or measured the dual effect of job stress on increasing or decreasing weight. Hence, we cannot conclude whereas this would influence or not the final result.

Clinical approach: There are many types of questionnaires used to evaluate the level of stress. Selecting the most suitable instrument for an organization is sometimes difficult and involvement of psychologists and human resources specialists might be of great help. In individual assessment, the detailed anamnesis, the working time and the perceived complexity of job demands are more easily revealed during the consultation. As in other occupational medicine consultations, attention should be paid to repetitive group complaints that underline a common organizational problem. Collaboration with the line managers and human resource management in a stress reduction programme is essential.

Health status also influences the employment choice and the occupational specialist has a role in modulating this choice. For example, fatigue is one of the reasons people quit shift work and organic causes should not be omitted before claiming stress as the causative factor. Clinical examination and exchange of medical information with the general practitioner are beneficial in finding the best possible solution for the employee.

3.4. Chemical hazards and air pollution

Assessing the health impact of toxic exposure is an important part of the occupational medicine practice. During the past decades, industrial exposure to chemicals has been gradually reduced in many countries by better safety measures, but low level constant and long-term exposure are still a matter of concern for many industrial and service workplaces. Occupational exposure to heavy metals occurs in many industries not only in the metal industry (mainly from electric air furnace or welding), metal refineries, battery production, colorant and pigment industries, smelter operations, radiator repair shops, mining, construction, but also in the modern days in the recycling industry and copying machine chambers. Indoor pollution and environmental exposure (from drinking water, outdoor pollution) are other possible sources of particles and chemical exposure linked to insulin resistance.

Heavy metals: Numerous studies indicated an association between heavy metals and insulin resistance (cardio-vascular diseases and diabetes) [122–127] explained by their direct effects on ROS production and inflammatory mediated cellular and systemic damage, decreased mitochondrial function with reduction of the capacity for fatty acid oxidation, inhibition of lipoprotein lipase in adipocytes and β-cells destruction [128]. Maternal gestational diabetes is also increased by heavy metal exposure [129].

Arsenic is well known for its carcinogenic, skin and neurological action. Arsenic epidemiological data about diabetes and cardio-vascular effects is quite extensive and comes from very different countries, both developed and under developed ones [130, 131]. Industrial activity and drinking water are the major source of exposure [132]. In drinking water, levels higher than 150 mg/L are positively associated with diabetes. Lower drinking water levels gave contradictory results [133, 134] in different studies, most probably needing a longer term exposure to produce the negative health effects. Arsenic is a highly reactive metal. Inside hepatocytes, the methylation of arsenate and arsenite generates monomethylarsonate (MMA) and dimethylarsinate (DMA), the latter being less reactive and more rapidly eliminated in urine. Metabolomics techniques revealed 103 urinary and 32 plasma metabolites associated with the urinary arsenic level. Interestingly, metabolites related to energy (ATP) production were common in diabetes and non-diabetes individuals, although the energy generation is different at the cellular level, insulin is active or if there is an IR. There were also other compounds affecting the citric acid cycle or the amino acid metabolism that were specifically modified only in diabetes, supporting the influence of arsenic on the diabetes status [135]. Under moderate environmental arsenic exposure [136], high inorganic urinary As and urinary DMA excretion were associated with high diabetes incidence; higher MMA had an inverse effect. Experimental data also supports an action of DMA on adipocytes (reduction of the insulin signal) and on beta cells apoptosis. Exposure burden, the type of arsenic compounds but also polymorphisms of arsenic (III) methyltransferase gene (*AS3MT*) [137] are relevant for the relation between MMA and DMA and nutrition, particularly on folate intake.

In animal experiments, arsenic has many pathophysiological effects related to insulin resistance: it reduces the intracellular insulin signal by suppressing the expression and phosphorylation of protein kinase B (PKB/Akt) and by direct inhibition of the Akt pathway [138]. Arsenic affects hepatic glucose output, increasing gluconeogenesis and up-regulates genes involved in oxidative and inflammation processes. As other heavy metals, arsenic causes mitochondrial toxicity reflected by the decreased level of the mitochondrial deacetylase, of FOXO 3a, ionic transport and membrane potential, decreased activity of manganese superoxide dismutase and increased reactive oxygen species [139]. Arsenic activates stress sensitive signalling pathways initiating pancreatic cell apoptosis [140] and oxidative stress in liver cells [141]. Arsenic has also a pro-thrombotic action and acceleration of atherosclerosis explaining the cardio-vascular effects [142, 143].

Mercury (Hg): Cross-sectional and prospective studies showing increases prevalence of diabetes [144, 145] in people exposed to Hg, which are not confirmed by others [146]. Hg is a very reactive metal: it interacts with the -SH groups of different enzymes, antioxidants (glutathione), and even with amino acids. Substantial experimental data concluded that Hg has both direct pro-oxidant effects and a capacity to reduce the antioxidant defence mechanism [147]. These characteristics have been confirmed in high occupational exposure [148, 149]. Hg reduces gluthatione, superoxide dismutase, catalase and thioredoxin reductase [150]. Denaturation of proteins increases the endoplasmic reticulum stress, initiating apoptosis. The inflammatory status is materialized by the increased levels of IL4, IL6, IL7 and TNFα and IFNγ [150]. Apart from these systemic effects, Hg has also more specific effects related to IR development: Hg has a direct toxic, pro-apoptotic effect on β-cells [151]; it decreases the insulin effects [152] and induces apoptosis in adipocytes.

Other heavy metals have also been involved in metabolic inflammation. During the cellular immune response, it was demonstrated that radicals induced by lead (Pb) disrupt the transcription signalling pathway mediated by the mitogen-activated protein kinase, nickel (Ni) induces the NLRP3-ASC-caspase 1 and As induces tyrosine kinase Src. The nuclear factor kB is activated by Pb, Ni or Hg with an increase in gene transcription for early inflammatory cytokines, such as Interleukin 1, interleukin 6 and tumour necrosis factor. Some metals, such as cadmium (Cd), can activate an inflammatory response through tissue damage induction mediated by free radicals, which results in leukocyte recruitment and cytokine secretions. Inflammation generated by metals can be reduced by metallothionein that has the ability to scavenge free radicals and bind toxic metals through the release of Zn and oxidation of SH groups [153].

From my previous personal clinical experience, in the Occupational Medicine Clinic of the Colentina University Hospital in Bucharest, I can confirm that workers from a lead battery factory, currently closed, admitted with saturnine colic, that had heavy Pb exposure, presented almost constantly a reversible hypertension syndrome.

Heavy metals and smoking: Smoking habit is one of the most important sources of cadmium in the general population. In fact, reduction of cadmium levels in the general population has been attributed to the efficiency of the anti-tobacco campaigns and policies [154]. Lead, nickel and chromium are also found in the tobacco smoke at significant levels. The relationship between smoking and diabetes is grounded on large epidemiological studies [155, 156]. Mechanisms such as impaired peripheral glucose uptake, oxidative stress, metabolic inflammation [157], and retention of the inhaled high concentration of metals (As, Al, Cd, Ni, Hg and Pb) [158, 159] have been described. A synergic effect of occupational exposure and smoking was found in smelters [160] or mining workers [159]; the common proposed pathological mechanism is the reduction of the anti-oxidative defence [160] or the increased retention of the heavy metals.

Other toxicants: Epidemiological studies brought the attention on other environmental toxicants such as persistent organic pollutants (POP), endocrine disruptors, or substances involved in the toxic oil syndrome (presumably aniline). POP is a generic term for a variety of chloride or bromide compounds, including pesticides and dioxin. Longitudinal studies showed that persons with high basal level of blood POP are at increased risk for diabetes. Studies that have also monitored the PCB were able to exclude the non-reverse effect of diabetes [161]. The stronger evidence of these associations came from occupational exposure studies: high accidental environment contamination [162] increased the relative risk of diabetes and of cardiovascular disease. Non-accidental occupational exposure in cohorts from 12 different countries also confirmed the relationship [163]. Bisphenol A is also a POP, well known for its endocrine disruptor effects; besides other negative effects, the rate of gestational diabetes is increased in women with a higher blood level [164]. Toxic oil syndrome-registered patients have threefold increase in prevalence of diabetes, and higher incidence of metabolic syndrome [165, 166].

The intimate mechanism of these associations is not completely found and most probably has certain specificities according to the particular substance. Many pathological pathways have been implicated: the increased accumulation of abdominal fat [167], inflammatory markers [168], adipocyte cytokines, TNFα and nuclear transcription factor kappa B (NFκB) [47], uric acid [169]. As a direct effect on the insulin signal, in animal experiments, perfluorooctane sulfonate inhibits the insulin Akt, signal leading to insulin resistance [170].

Particulate exposure: Air pollution, particularly fine and ultrafine particles, has been related to the acute incidence of cardio-vascular events and asthma attacks from inflammatory, oxidative mechanisms and pro-thrombotic mechanisms activation [171]. This correlation remained consistent after exclusion of possible confounders (education level, work exposure, alcohol, vegetable and fruit consumption, smoking or physical activity) [172]. Initially described for cardiovascular effects [171, 173], the particulate exposure includes now also the increased risk for diabetes [174, 175] and for non-alcoholic steatohepatitis [176]. Particulate matter (PM) is generally a subject of environmental health, and exposure during the day is very variable depending the microenvironment (home, kitchen, policy of smoking indoor, traffic, general level of city pollution control, occupation) [177]. However, there are also many workplaces were PM10 (the respirable component), PM2.5 (fine particles) and the PM0.1 (ultrafine particles) have still high values. Road transport, energy production and distribution, waste as well as domestic exposure (cooking) are major sources in Europe [178]. Photocopiers centres, sick buildings might also have PM10 and PM2.5 above than permissible levels.

Large cohort studies in North America showed that a 10 g/mm^3 increase in PM2.5 was associated with 13% increases in hypertension and with 18% increase in diabetes-related mortality, concluding that current PM 2.5 US standards are too high for the long-term exposure and should be reduced [179].

In different countries or regions, the proportion and the composition of air pollution differ; in a China study, fine particles bound polycyclic aromatic hydrocarbons (PAH) and this characteristic might have a cumulative effect on IR, as it has independently associated with this risk [180]. PAH is an environmental pollutant but exposure is increased in occupations that involve bitumen and asphalt manipulation, in foundries, distilleries, painting industry and other manufactories [181, 182]. Firefighters are also exposed to significant levels. Smoking is another important PAH source and, therefore, efforts to reduce this habit in workers occupationally exposed are highly recommended.

The particulate matters IR pathophysiologic mechanisms are rather similar to the ones involved in heavy metal exposure: obesity risk increase [183], enhancement of inflammation, with higher plasma levels IL8, ICAM-1 or eosinophilic cationic protein [184], endoplasmic reticulum stress and insulin signalling interference in a dose-response manner [172]. Reduction in antioxidant defence is supported by both experimental on cell lines and from the increased effect on persons with GSTM1 null alleles [184]. Disruption of metabolic genes has also been reported [185]. Several studies proposed a link between genetic variants of ABO locus; particulate exposure and IR risk [186]. When occupational exposure to particles (particularly to asbestos, welding fumes, respiratory irritants) was added to the epidemiological model, the risk of group O subjects was enhanced [187]. There is no conclusive explanation about this association, but ABO antigens are expressed also on the von Willebrand factor, presumably increasing the prothrombotic and chronic inflammation risk [188] and the ABO locus is related to the E-selectin, ICAM-1 and TNF-α blood level [186].

Cardio-vascular and ECG modifications: Controlled particulate exposure showed high systolic [189] (or high systolic and diastolic increase [172] according to the type of particle), decreased of blood plasminogen and thrombomodulin, and an increase in the inflammatory markers: CRP and serum amyloid [190]. Diesel exhaust double-blind experiment mean peak effect on systolic blood pressure increases between 30 and 60 min. after exposure and remained high 24 h after, independent of metabolic syndrome previously existing components. Interestingly, in this experiment, the increase in blood pressure was not related to perception of exposure [189]. Persons with metabolic syndrome exposed to ultrafine particles and GSTM1 null allele are particularly at risk of altered repolarization and increased QT duration [191].

Clinical approach: Indoor air pollution measurement is part of the occupational safety procedures and whenever data are available there should be transmitted to the occupational physician. The detailed occupational anamnesis, symptoms, clinical examination and existing exposure indicators orient the medical conduct.

For heavy metals exposure, biological monitoring is recommended in workplaces or occupations at risk. The biological monitoring is included in the national recommendations for occupational medicine practice in many countries. The limit values of the exposure indicators are defined for chronic poisoning. On what concerns the metabolic risk, there is no threshold limit established yet and general population data suggest that lower limits than those for chronic poisoning would be a reasonable approach. We cannot conclude this topic, until well-designed studies to address this specific objective are conducted. On the other hand, until these studies are available and taken into consideration the increasing incidence of insulin resistance worldwide, efforts to reduce this non-classical risk factor should not be postponed.

A special concern is related to the maternal and foetal risk and to the gestational diabetes. Ideally, pregnant women should not be exposed to chemicals; if environmental contamination is not easy to avoid, at least the occupational one should be limited. Besides the negative effects of chemicals on the foetus development, the increased risk of gestational diabetes adds unfavourable short- and long-term consequences.

4. Workplace interventions

As insulin resistance implication is better understood, and its prevalence increases worldwide, the number of interventions that have been published in last decade has also increased. Most of them, even if they directly target insulin resistance are better known as workplace wellness programmes. In countries were private insurance is paid by the employer, they could be a mean to reduce the insurance plans; in others, where work-promotion activities are more and more valued by the companies, they are conducted either as a channel to encourage individuals to take preventive measures through education and screening or as employment-based activity to promote health behaviours and disease management, such as stress management, adjustable work station to reduce continuous sitting, team support groups and cafeteria/canteen provision of healthy food initiatives. Although very diverse, there are however some general characteristics of these interventions:

1. *Assessment of the health risk according to the working conditions*. Risk assessment is always the first step as the intervention has to be meaningful for the organization and should reflect the specific issues of the employees. Working conditions and health-related habits should be analysed, as multi-component interventions have given better results [160]. Sometimes, just screening programmes initiated inside the company increase awareness and initiate individual actions [161]. In works with high aerobic workload, a low cardiovascular fitness is a physical risk factor and a stressor. Increasing the aerobic fitness by initiating an exercise training programme was able to decrease the basal heart rate, the BMI and the inflammatory markers (high sensitivity protein C), to increase work efficiency and to reduce stress related to work overload. Sedentary behaviour has benefited from different ergonomically interventions [162, 163]. Some studies have emphasized the benefit of sitting interruption during the working day; efficiency of short 1–2 min every half hour of sitting break was better in terms of metabolic syndrome criteria compared to the two 15 min breaks per workday [164]. In certain circumstances, cumulative risk factors are encountered: for example, seafarers are recognized as a profession with high incidence of metabolic syndrome. An intervention including healthy cooking training for ship cooks together with improvement of the ships fitness facilities and education (anti-smoking, individual exercise, extra health checks) led to a decrease in sugar intake, increased physical fitness scores and a significantly decrease in prevalence of the metabolic syndrome [165, 166].

2. *Obtaining the maximum support of the management team*. These interventions are changing behaviour interventions; they imply not only financial cost but also time allocation, organizational changes, etc. If management is not convinced and does not actively supports the programme, making it part of the organizational culture, the chance of achieving the proposed results is considerably reduced. Presenteeism, defined as being present at work, but with a limited performance in some aspects of job related a health problem, is often a high hidden cost for employers. Improving the general health status and well-being of their employees could be, therefore, perceived as a valuable investment. Stress management programmes, in particular, need specific organizational behaviour assessment and

a proactive participation of the line and human resource management. Of course, there are some universally interventions for mental health [167], but they should be aligned with other managerial and organizational initiatives [168].

3. *Proper preparation and communication inside the company*: Ideally, some representatives of the beneficiaries should be involved in preparation of the intervention, for presenting their needs and for choosing from different possibilities of implementation the best suited for them. Communication of the aim, the means and the expected results is necessary before, during and after the intervention. It is reasonable to consider that employees are in different stages of motivation for change. Therefore, stage-based intervention for non-intenders, intenders and actors (people who have already taken action towards changing behaviour) could be more effective [169, 170]. Barriers to adherence and completion of the programme are better solved in a participatory approach [171].

4. *Adapted tools*: There is no one solution for all organizations. For some, the online support is possible and is part of their working culture; in other organizations, group or individual discussions are more effective. Computer-based interventions have a better reach, particularly in larger organizations. A Cochrane systematic review on interactive computer-based interventions for weight loss or weight maintenance in overweight or obese people found that such interventions significantly reduced body weight [172]. Combined systems with face-to-face and computer/telephone interactions are also used.

5. *Team work*, from the first steps of design until results evaluation. Nobody can expect that a transformational change can be proposed and delivered solely by the occupational physician, but it is desirable that he is an active member of the team [173]; psychologists, ergonomists, nutritionists, exercise trainers, IT and communication specialists, etc. are needed, according to the scope of the intervention. Concerning risk reduction of the metabolic risk, nutrition is a traditional risk factor for insulin resistance and trained dieticians should be involved. Nutrition knowledge make people less likely to be hypertensive compared to one with low level of healthy nutrition [174] and when combined with a physical activity programmes [175], and facilitated access to healthy food, become more effective [176]. Workplaces are excellent settings for health promotion programmes, joining all these conditions together, if there is managerial commitment and support.

5. Conclusions

The minimum requirements for a health issue to become an occupational medicine subject are a proven relation with the working conditions, and a benefit from a working place intervention. This chapter had provided arguments that insulin resistance has both characteristics. It has, however, a specificity, that in some countries had brought benefits for including it in the occupational services and in others have been an obstacle: insulin resistance shares both elements of the classical health protection and of the health promotion activities.

In terms of health protection, insulin resistance is related to medium or even low exposure to air pollution (particles, solvents) or heavy metals. The biological monitoring reference for insulin resistance is not yet defined, making difficulties in interpreting field data. Using the general population assessments might be a solution, but there is clearly a need for well-designed research in this area to reach a valid conclusion. Insulin resistance is also related to other work hazards: work schedules, sedentary behaviour and stress. Concerning the work schedules, we benefit of many data, experience and general recommendations about working on shifts; flexible working time is a new challenge and probably needs more individual solutions. Occupational stress is recognized as one of the most important risk factor of the modern society and, even if it is more in the occupational psychology field, it has to be integrated in the occupational medicine service, as many other medical conditions, except insulin resistance, are aggravated by stress.

In terms of health promotion, prevention of insulin resistance and its major consequences (the cardio-vascular disease and type-2 diabetes) are perceived as part of the healthy life-style and well-being programmes. Healthy nutrition, smoking cessation, physical activity and sleep hygiene are the pillars of these programmes. The appropriate and culturally adapted design increases the adherence to these programmes; their real success is the integration of the healthy behaviours and the reduction of obesity and insulin resistance-related mortality. Management support is needed in the preliminary phases (approving and promoting the programme) but also during the implementation phase, as many of these activities request working condition adaptations. Better employee engagement is obtained by involving them from the early steps of the programme.

Therefore, assessment of the risk, screening and workplace intervention programmes to reduce insulin resistance incidence are part of the occupational medicine activity and should be included in the occupational health service provision. Further research to develop tools for on appropriate individual and organizational assessment and strategies for interventions is needed.

Author details

Marina Oțelea

Address all correspondence to: dr.marinaotelea@gmail.com

Clinical Department 2, University of Medicine and Pharmacy "Carol Davila", Bucharest, Romania

References

[1] WHO. Media centre. Cardio-vascular diseases [Internet]. Available from: http://www.who.int/mediacentre/factsheets/fs317/en/ [Accessed 15.05.2016].

[2] WHO Media centre. Diabetes [Internet]. Available from: http://www.who.int/mediacentre/factsheets/fs312/en/ [Accessed 15.05.2016].

[3] Winkleby MA, Jatulis DE, Frank E, Fortmann SP. Socioeconomic status and health: How education, income, and occupation contribute to risk factors for cardiovascular disease. Am J Pub Health. 1992;**82**:816–820.

[4] Santos A, Ebrahim S, Barros H. Gender, socio-economic status and metabolic syndrome in middle-aged and old adults. BMC Public Health. 2008;**8**:62. DOI: 10.1186/1471-2458-8-62.

[5] Al-Daghri NM, Alkharfy KM, Al-Attas OS, Khan N, Alfawaz HA, Alghanim SA, et al. Gender-dependent associations between socioeconomic status and metabolic syndrome: A cross-sectional study in the adult Saudi population. BMC Cardiovasc Disord. 2014;**14**:51. DOI: 10.1186/1471-2261-14-51.

[6] Langenberg C, Kuh D, Wadsworth MEJ, Brunner E, Hardy R. Social circumstances and education: Life course origins of social inequalities in metabolic risk in a prospective national birth cohort. Am J Pub Health. 2006;**96**:2216–2221. doi:10.2105/AJPH.2004.049429.

[7] Schooling CM, Jiang CQ, Lam TH, Zhang WS, Cheng KK, Leung G. M. Life-course origins of social inequalities in metabolic risk in the population of a developing country. Am J Epidemiol. 2008;**167**:419–428.

[8] Gupta R, Deedwania PC, Sharma K, Gupta A, Guptha S, Achari V et al. Association of educational, occupational and socioeconomic status with cardiovascular risk factors in Asian Indians: A cross-sectional study. Gupta V. PLoS ONE. 2012;**7**:e44098. DOI:10.1371/journal.pone.0044098.

[9] Lucove JC, Kaufman JS, James SA. Association between adult and childhood socioeconomic status and prevalence of the metabolic syndrome in African Americans: The pitt county study. Am J Public Health. 2007;**97**:234–236. DOI:10.2105/AJPH.2006.087429.

[10] Eibner C, Evans WN. Relative deprivation, poor health habits, and mortality. J Hum Res, 2005:**40**:591–620.

[11] Popkin BM, Gordon-Larse P. The nutrition transition: worldwide obesity dynamics and their determinants. Int J Obes Relat Metab Disord. 2004;**28**:S2–S9. DOI:10.1038/sj.ijo.0802804.

[12] Omran AR. The epidemiologic transition: A theory of the epidemiology of population change. The Milbank Q. 2005;**83**:731–757. DOI:10.1111/j.1468-0009.2005.00398.x.

[13] Cornman JC, Glei DA, Goldman N, Ryff CD, Weinstein M. Socioeconomic status and biological markers of health: An examination of adults in the United States and Taiwan. J Aging Health. 2015;**27**:75–102.

[14] Dallongeville J, Cottel D, Ferrières J, Arveiler D, Bingham B, Ruidavets JB, et al. Household income is associated with the risk of metabolic syndrome in a sex-specific manner. Diabetes Care. 2005;**28**:409–415. DOI: 10.2337/ diacare.28.2.40.

[15] Davila EP, Florez H, Fleming LE, et al. Prevalence of the metabolic syndrome among U.S. workers. Diabetes Care. 2010;**33**:2390–2395. DOI:10.2337/dc10-0681.

[16] Alegría E, Cordero A, Laclaustra M, Grima A, León M,. Casasnovas J. Prevalence of metabolic syndrome in the Spanish working population: MESYAS registry. Rev Esp Cardiol. 2005;**58**:797–806.

[17] Lee W, Yeom H, Yoon JH, Jung PK, Lee JH, Seok H, et al. Metabolic outcomes of workers according to the International Standard Classification of Occupations in Korea. Am J Ind Med. 2016;**59**:685–694.

[18] Church TS, Thomas DM, Tudor-Locke C, Katzmarzyk PT, Earnest CP, Rodarte RQ. Trends over 5 decades in U.S. occupation-related physical activity and their associations with obesity. PLoS ONE. 2011;**6**:e19657.

[19] Bertrais S, Beyeme-Ondoua, JP, Czernichow S, Galan P, Hercberg S, Oppert JM. Sedentary behaviors, physical activity, and metabolic syndrome in middle-aged French subjects. Obes Res, 2005;**13**:936–944. DOI:10.1038/oby.2005.108.

[20] Herman MA, Kahn BB. Glucose transport and sensing in the maintenance of glucose homeostasis and metabolic harmony. JCI. 2006;**116**(7):1767–1775.

[21] Poehlman ET, Dvorak RV, DeNino WF, Brochu M, Ades PA. Effects of resistance training and endurance training on insulin sensitivity in nonobese, young women: A controlled randomized trial. J Clin Endocrinol Metab. 2000;**85**:2463–2468. http://doi.org/10.1210/jcem.85.7.6692.

[22] Menshikova EV, Ritov VB, Fairfull L, Ferrell RE, Kelley DE, Goodpaster BH. Effects of exercise on mitochondrial content and function in aging human skeletal muscle. J Gerontol A, Biol Sci Med Sci. 2006;**61**:534–540.

[23] Balakrishnan VS, Rao M, Menon V, et al. Resistance training increases muscle mitochondrial biogenesis in patients with chronic kidney disease. Clin J Am Soc Nephrol. 2010;**5**:996–1002.

[24] Melov S, Tarnopolsky MA, Beckman K, Felkey K, Hubbard A. Resistance exercise reverses aging in human skeletal muscle. PLoS One. 2007;**2**:e465.

[25] Giresi P, Stevenson E, Theilhaber J, Koncarevic A, Parkington J, Fielding R, et al. Identification of a molecular signature of sarcopenia. Physiol Genomics 2005;**21**:253–263.

[26] Yin H, Price F, Rudnicki MA. Satellite cells and the muscle stem cell niche. Physiol Rev. 2013;**93**:23–67. DOI:10.1152/physrev.00043.2011.

[27] Civitarese AE, Carling S, Heilbronn LK, Hulver MH, Ukropcova B, Deutsch WA. Calorie restriction increases muscle mitochondrial biogenesis in healthy humans. PLoS Medicine. 2007;**4**:485–494.

[28] Laaksonen DE, Lakka HM, Lynch J, Lakka TA, Niskanen L, Rauramaa R, et al. Cardiorespiratory fitness and vigorous leisure-time physical activity modify the association of small size at birth with the metabolic syndrome. Diabetes Care. 2003;**26**:2156–2164.

[29] Randle PJ, Garland PB, Hales CN, Newsholme EA. The glucose fatty-acid cycle. Its role in insulin sensitivity and the metabolic disturbances of diabetes mellitus. Lancet. 1963;**1**:785–789.

[30] Wells J, Siervo M. Obesity and energy balance: Is the tail wagging the dog? Eur J Clin Nutr. 2011;**65**:1173–1189. DOI:10.1038/ejcn.2011.132.

[31] Cunningham JJ. Body composition as a determinant of energy expenditure: A synthetic review and a proposed general prediction equation. Am J Clin Nutr. 1991;**54**:963–969.

[32] Johnstone AM, Murison SD, Duncan JS, Rance KA, Speakman JR. Factors influencing variation in basal metabolic rate include fat-free mass, fat mass, age, and circulating thyroxine but not sex, circulating leptin, or triiodothyronine. Am J Clin Nutr. 2005;**82**:941–948.

[33] Dvorak RV, DeNino WF, Ades PA, Poehlman ET. Phenotypic characteristics associated with insulin resistance in metabolically obese but normal weight young women. Diabetes.1999;**48**:2210–2214.

[34] Bays HE. Adiposopathy. Is "sick fat" a cardiovascular disease? J Am Coll Cardiol. 2011;**57**:2461–2473.

[35] Sisson SB, Camhi SM, Church TS, Martin CK, Tudor-Locke C, Bouchard C, et al. Leisure time sedentary behavior, occupational/domestic physical activity, and metabolic syndrome in U.S. men and women. Metab Syndr Relat Disord. 2009;**7**:529–536.

[36] Pietroiusti A, Magrini A, Coppeta L, Barone S, Sed B, Manuli G, et al. Metabolic syndrome among operators using video display units in call centers. Scand J Work, Environ Health. 2007;**Suppl 33(3)**:49–53.

[37] Shin SY, Lee CG, Song HS, Kim SH, Lee HS, Jung MS, et al. Cardiovascular disease risk of bus drivers in a city of Korea. Ann Occup Environ Med. 2013;**25**:34.

[38] Leischik R, Foshag P, Strauss M, Littwitz H, Garg P, Dworrak B, et al. Aerobic capacity, physical activity and metabolic risk factors in firefighters compared with police officers and sedentary clerks. PLoS ONE. 2015;**10**: e0133113. DOI:10.1371/journal.pone.0133113.

[39] Parry S, Straker L. The contribution of office work to sedentary behaviour associated risk. BMC Public Health. 2013;**13**:1–10. http://doi.org/10.1186/1471-2458-13-296.

[40] Healy G, Dunstan DW, Salmon J, Cerin E, Shaw J, Zimmet P, et al. Breaks in sedentary time. Beneficial associations with metabolic risk. Diabetes Care. 2008;**31**:661–666.

[41] Bey L, Hamilton MT. Suppression of skeletal muscle lipoprotein lipase activity during physical inactivity: A molecular reason to maintain daily low-intensity activity. J Physiol. 2003;**551**:673–682.

[42] Hamilton M, Hamilton D, Zderic T. Role of Low Energy Expenditure and Sitting in Obesity, Metabolic Syndrome, Type 2 Diabetes, and Cardiovascular Disease. Diabetes. 2007;**56**:2655–2667.

[43] Mabry RM, Winkler EH, Reeves MM, Eakin EG, Owen N. Associations of physical activity and sitting time with the metabolic syndrome among Omani adults. Obesity. 2012;**20**:2290–2295.

[44] Stiller JW, Postolache TT. Sleep-wake and other biological rhythms: Functional neuro-anatomy. Clin Sports Med. 2005;**24**:205–235.

[45] Wong SN, Halaki M, Chow CM. The periodicity of sleep duration —an infradian rhythm in spontaneous living. Nat Sci Sleep. 2013;**5**:1–6.

[46] Toh KL. Basic science review on circadian rhythm biology and circadian sleep disorders. Ann Acad Med Singapore. 2008;**37**:662–668.

[47] Nakagawa H, Sack RL, Lewy AJ. Sleep propensity free-runs with the temperature, melatonin, and cortisol rhythms in a totally blind person. Sleep. 1992;**15**:330–336.

[48] Reiter RJ, Tan DX, Mayo JC, Sainz RM, Leon J, Czarnocki Z. Melatonin as an antioxidant: Biochemical mechanisms and pathophysiological implications. Acta Biochim Pol. 2003;**50**:1129–1146.

[49] Hense S, Barba G, Pohlabeln H, De henauw S, Marild S, Molnar D et al. Factors that influence sleep duration in European children. Sleep. 2011;**34**(5):633–639.

[50] Leng Y, Wainwright NWJ, Cappuccio FP, Surtees PG, Luben R, Wareham Net al. Self-reported sleep patterns in a British population cohort. Sleep Med. 2014;**15**:295–302. DOI:10.1016/j.sleep.2013.10.015.

[51] Hao YL, Zhang B, Jia FJ, Li XL, Tang Y, Ren YZ et al. A three-phase epidemiological study of short and long sleepers in a middle-aged Chinese population: Prevalence and characteristics. Braz J Med Biol Res. 2014;**47**:157–165.

[52] Touchette E, Côté SM, Petit D, Liu X, Boivin M, Falissard B, et al. Short nighttime sleep-duration and hyperactivity trajectories in early childhood. Pediatrics. 2009;**124**:e985.

[53] He Y, Jones CR, Fujiki N, Xu Y, Guo B, Holder JL Jr, et al. The transcriptional repressor DEC2 regulates sleep length in mammals. Science (NY). 2009;**325**:866–870. DOI:10.1126/science.1174443.

[54] Lakerveld J, Mackenbach JD, Horvath E, Rutters F, Compernolle S, Bárdos H, et al. The relation between sleep duration and sedentary behaviours in European adults. Obes Rev. 2016;**17**:62–67.

[55] Archer SN, Laing EE, Möller-Levet CS, van der Veen DR, Bucca G, Lazar AS, et al. Mistimed sleep disrupts circadian regulation of the human transcriptome. Proc Natl Acad Sci USA. 2014;**111**:E682–E691.

[56] Partinen M, KaprioJ, Koskenvuo M, Putkonen P, Langinvainio H. Genetic and environmental determination of human sleep. Sleep. 1983;**6**:179–185.

[57] Maurovich-Horvat E, Pollmächer T, Šonka K. The effects of sleep and sleep deprivation on metabolic, endocrine and immune parameters. Prague Med Rep. 2008;**109**:275–285.

[58] Peschke E, Frese T, Chankiewitz E, Peschke D, Preis U, Schneyer U, et al. Diabetic Goto Kakizaki rats as well as type 2 diabetic patients show a decreased diurnal serum melatonin level and an increased pancreatic melatonin-receptor status. J Pineal Res. 2006;**40**:135–143.

[59] Frellick F. Melatonin links gene variant to Type 2 diabetes. 2016. Available from: http://www.medscape.com/viewarticle/863531 [Accessed 2016-08-25]

[60] Engle-Friedman M. The effects of sleep loss on capacity and effort. Sleep Science. 2014;7:213–224. http://doi.org/10.1016/j.slsci.2014.11.001.

[61] Schmid SM, Hallschmid M, Jauch-Chara K, Bandorf N, Born J, Schultes B. Sleep loss alters basal metabolic hormone secretion and modulates the dynamic counterregulatory response to hypoglycemia. J Clin Endocrinol Metab. 2007;**92**:3044–3051. http://doi.org/10.1210/jc.2006-2788

[62] Broussard JL, Ehrmann DA, Van Cauter E, Tasali E, Brady MJ. Impaired insulin signaling in human adipocytes after experimental sleep restriction: A randomized, crossover study. Ann Intern Med. 2012;**157**:549–557.

[63] Mander BA, Colecchia E, Spiegel KS, Van Cauter E. Short Sleep: A risk factor for insulin resistance and obesity. Sleep. 2001;**Suppl 24**:A74–75.

[64] Leproult R, Deliens G, Gilson M, Peigneux P. Beneficial impact of sleep extension on fasting insulin sensitivity in adults with habitual sleep restriction. Sleep. 2015;**38**:707–715. DOI:10.5665/sleep.4660.

[65] Hampton SM, Morgan LM, Lawrence N, et al. Postprandial hormone and metabolic responses in simulated shift work. J Endocrinol. 1996;**151**:259–267.

[66] Spiegel K, Tasali E, Penev P, van Cauter E. Brief communication: Sleep curtailment in healthy young men is associated with decreased leptin levels, elevated ghrelin levels and increased hunger and appetite. Ann Intern Med, 2004;**141**:846–850.

[67] Canuto R, Pattussi MP, Macagnan JBA, Henn RL, Olinto MTA. Metabolic syndrome in fixed-shift workers. Rev Saúde Pública. 2015;**49**. http://doi.org/10.1590/S0034-8910.2015049005524.

[68] Scheer FA, Hilton MF, Mantzoros CS, Shea SA. Adverse metabolic and cardiovascular consequences of circadian misalignment. PNAS. 2009;**106**:4453–4458.

[69] Litinski M, Scheer FA, Shea SA. Influence of the circadian system on disease severity. Sleep Med Clin. 2009;**4**:14v3–163. DOI:10.1016/j.jsmc.2009.02.005.

[70] Garaulet M, Ordovás JM, Madrid JA. The chronobiology, etiology and pathophysiology of obesity. Int J Obes. 2010;**34**:1667–1683.

[71] Schoeller D. Entrainment of the diurnal rhythm of plasma leptin to meal timing. J Clin Invest. 1997;**100**:1882–1887.

[72] Marcheva B, Ramsey KM, Buhr ED, Kobayashi Y, Su H, Ko CH et al. DiAQ4sruption of the clock components CLOCK and BMAL1 leads to hypoinsulinaemia and diabetes. Nature 2010:**466**:627–663.

[73] Maury E, Ramsey M, Bass KJ. Circadian rhythms and metabolic syndrome: From experimental genetics to human disease. Circ Res. 2010;**106**(3):447–462.

[74] Lehman JJ, Brager PM, Kovacs A, Saffitz JE, Medeiros DM, Kelly DP. PPARγ coactivator-1 (PGC-1) promotes cardiac mitochondrial biogenesis. J Clin. Invest. 2000;**106**:847–856.

[75] Taheri S, Lin L, Austin D, Young T, Mignot E. Short sleep duration is associated with reduced leptin, elevated ghrelin, and increased body mass index. PLoS Medicine. 2004;**1**:e62.

[76] Imaki M, Hatanaka Y, Ogawa Y, Yoshida Y, Tanada S. An epidemiological study on relationship between the hours of sleep and life style factors in Japanese factory workers. J Physiol Anthropol Appl Human Sci. 2002;**21**:115–120.

[77] Oswald I, Taylor AM, Treisman M. Discriminative responses to stimulation during human sleep. Brain. 1960;**83**:440–453.

[78] Horne J, Östberg O. A self-assessment questionnaire to determine morningness eveningness in human circadian rhythms. Int J Chronobiol. 1976;**4**:97–100.

[79] Doghramji K, Mitler MM, Sangal R, Shapiro C, Taylor S, Walsleben J, et al. A normative study of the maintenance of wakefulness test (MWT). Electroencephalogr Clin Neurophysiol. 1997;**103**:554–562.

[80] Brunner EJ, Hemingway H, Walker BR, Page M, Clarke P, Juneja M, et al. Adrenocortical, autonomic, and inflammatory causes of the metabolic syndrome, nested case-control study. Circulation. 2002;**106**:2659–2665.

[81] Wilbert-Lampen U, Nickel T, Leisner D, Guthlin D, Matis T, Volker C, et al. Modified serum profiles of inflammatory and vasoconstrictive factors in patients with emotional stress-induced acute coronary syndrome during world cup soccer 2006. J Am Coll Cardiol. 2010;**55**:637–642.

[82] Kouvonen A, Kivimäki M, Virtanen M, Heponiemi T, Elovainio M, Pentti J, et al. Effort-reward imbalance at work and the co-occurrence of lifestyle risk factors: Cross-sectional survey in a sample of 36,127 public sector employees. BMC Public Health. 2006;6:24. DOI:10.1186/1471-2458-6-24.

[83] European agency for safety and health at work. Facts 22. Work related stress. 2002. Available from: https://osha.europa.eu/en/tools-and-publications/publications/factsheets/22 [Accessed: 2016-05-15]

[84] Sauter S, Murphy L, Colligan M, Swanson N, Hurrell J, Scharf F, et al. Stress at work. Available from: http://www.cdc.gov/niosh/docs/99-101/.[Accessed: 2016-05-15]

[85] Karasek R, Brisson C, Kawakami N, Houtman I, Bongers P, Amick, B. The job content questionnaire (JCQ): An instrument for internationally comparative assessments of psychosocial job characteristics. J Occup Health Psychol. 1998;**3**:322–355.

[86] Siegrist J, Starke D, Chandola T, Godin I, Marmot M, Niedhammer I, et al. The measurement of effort-reward imbalance at work: European comparisons. Soc Sci Med. 2004;**58**:1483–1499.

[87] Pelfrene E, Clays E, Moreau M, Mak R, Vlerick P, Kornitzer M, De Backer G. The job content questionnaire: Methodological considerations and challenges for future research by. Arch Public Health. 2003:**61**:53–74.

[88] Bakker AB, Demerouti E. The job demands resources model: State of the art. journal of managerial psychology. 2007;**22**:309–328. http://doi.org/10.1108/02683940710733115.

[89] Magnavita N. Work-related psychological injury is associated with metabolic syndrome components in apparently healthy workers. Plos One. 2015;**10**:e0130944.

[90] Rafferty AE, Griffin MA. Perceptions of organizational change: A stress and coping perspective. J Appl Psychol. 2006;**91**:1154–1162.

[91] Viitasalo K, Kuosma E, Laitinen J, Härmä M. Effects of shift rotation and the flexibility of a shift system on daytime alertness and cardiovascular risk factors. Scand J Work, Environ Health. 2008;**34**:198–205. http://doi.org/10.5271/sjweh.1228.

[92] Shields M. Long working hours and health. Health Rep. 1999;**11**:33–48.

[93] De Raeve L, Jansen NWH, Kant IJ. Health effects of transitions in work schedule, work hours and overtime in a prospective cohort study. Scand J Work, Environ Health. 2007;**33**:105–113.

[94] NIshiyama K, Johnson JV. Karoshi--death from overwork: Occupational health consequences of Japanese production management. Int J Helath Serv. 1997;**27**;625–641.

[95] Alvin M, Mellner C, Movitz F, Aronsson G. The diffusion of flexibility: Estimating the incidence of low-regulated working conditions. Nordic Journal of Working Life Studies. 2013;**3**:99–116.

[96] Janssen D, Nachreiner F. Health and psychosocial effects of flexible working hours. Rev Saude Publica. 2004;**Suppl 38**. http://dx.doi.org/10.1590/S0034-89102004000700003.

[97] Bergmann N, Gyntelberg F, Faber J. The appraisal of chronic stress and the development of the metabolic syndrome: A systematic review of prospective cohort studies. Endocr Connect. 2014;**3**:R55–R80. DOI:10.1530/EC-14-0031.

[98] Siegrist J. Effort—reward imbalance and health in a globalized economy. Scand J Work Envron Jealth. 2008;**suppl 6**:163–168.

[99] Joksimovic L, Starke D, van der Knesebeck O, Siegrist J. Perceived work stress, overcommitment, and self-reported musculoskeletal pain: A cross-sectional investigation. Int J Behav Med. 2002;**9**:122–138.

[100] Godin I, Kittel F. Differential economic stability and psychosocial stress at work: Associations with psychosomatic complaints and absenteeism. Soc Sci Med. 2004;**58**:1543–1553.

[101] De Jonge J, Bosma H, Peter R, Siegrist J. Job strain, effort-reward imbalance and employee well-being: A large-scale cross-sectional study. Soc Sci Med. 2000;**50**:1317–1327.

[102] Garbarino S, Magnavita N. Work stress and metabolic syndrome in police officers. A prospective study. PLoS ONE. 2015;**10**:e0144318. DOI:10.1371/journal.pone.0144318.

[103] Janczura M, Bochenek G, Nowobilski R, Dropinski J, Kotula-Horowitz K, Laskowicz B, et al. The relationship of metabolic syndrome with stress, coronary heart disease and pulmonary function—An occupational cohort-based study. PLoS ONE. 2015;10(8):1–20. http://doi.org/10.1371/journal.pone.0133750.

[104] Branth S, Ronquist G, Stridsberg M, Hambraeus L, Kindgren E, Olsson R, et al. Development of abdominal fat and incipient metabolic syndrome in young healthy men exposed to long term stress. Nutr Metab Cardiovasc Dis. 2007;**17**:427-435.

[105] Chandola T, Brunner E, Marmot M. Chronic stress at work and the metabolic syndrome: Prospective study. BMJ (Clinical Research Ed.). 2006;**332**:521–525.

[106] Price RH, Choi JN, Vinokur AD. Links in the chain of adversity following job loss: How financial strain and loss of personal control lead to depression, impaired functioning, and poor health. J Occup Health Psychol. 2002;**7**:302–312. http://doi.org/10.1037/1076-8998.7.4.302.

[107] Virtanen P, Vahtera J, Kivimaki M, Pentti J, Ferrie J. Employment security and health. J Epidemiol Community. 2002;**56**:569–574.

[108] Dragano N, Verde PE, Siegrist, J. Organisational downsizing and work stress: Testing synergistic health effects in employed men and women. J Epidemiol Community Health. 2005;**59**:694–699. http://doi.org/10.1136/jech.2005.035089.

[109] Pyykkonen AJ, Tuomi T, Eriksson JG, Groop L, Isomaa B. Stressful life events and the metabolic syndrome the prevalence, prediction and prevention of diabetes (PPP)-botnia study. Diabetes Care. 2010;**33**:378–384.

[110] Faragher EB, Cass M, Cooper CL. The relationship between job satisfaction and health: A meta-analysis. Occup Environ Med. 2005;**62**:105–112.

[111] Nichole Kelly. Associations between binge eating and executive functioning among young women. [thesis] Richmond VA, US: Virginia Commonwealth University. 2012.

[112] Dube L, LeBel JL, Lu J. Affect asymmetry and comfort food consumption. Physiol Behav. 2005;**86**:559–567. http://doi.org/10.1016/j.physbeh.2005.08.023.

[113] Drescher LS, Hasselbach J. Food Choices under Stress: Considering internet usage and social support. In: AAEA/EAAE/CAES joint symposium social networks, social media and the economics of food montréal, Québec. Canada, 29–30 May, 2014.

[114] Ree M, Riediger N, Moghadasian MH. Factors affecting food selection in Canadian population. Eur J Clin Nutr. 2008;**62**:1255–1262. http://doi.org/10.1038/sj.ejcn.1602863.

[115] Lampuré A, Schlich P, Deglaire A, Castetbon K, Péneau S, Hercberg S, et al. Sociodemographic, psychological, and lifestyle characteristics are associated with a liking for salty and sweet tastes in French adults. J Nutr. 2015;**145**:587–594.

[116] Camilleri GM, Mejean C, Kesse-Guyot E, Andreeva VA, Bellisle F, Hercberg S, et al. The associations between emotional eating and consumption of energy-dense snack foods are modified by sex and depressive symptomatology. J Nutr. 2014;**144**:1264–1273.

[117] Tan CC, Chow CM. Stress and emotional eating: The mediating role of eating dysregulation. Pers Indiv Differ. 2014;**66**:1–4. http://doi.org/10.1016/j.paid.2014.02.033.

[118] Oliver G, Wardle J, Gibson EL. Stress and food choice: A laboratory study. Psychosom Med 2000;**62**:853–865.

[119] Tan CC, Ruhl H, Chow CM, Ellis, L. Retrospective reports of parental feeding practices and emotional eating in adulthood: The role of food preoccupation. Appetite. 2016;**105**:410–415. DOI:10.1016/j.appet.2016.06.009.

[120] Eisenstein SA, Bischoff AN, Gredysa DM, Antenor-Dorsey JAV, Koller JM, Al-Lozi A, et al. Emotional eating phenotype is associated with central dopamine d2 receptor binding independent of body mass index. Sci Rep. 2015;**5**:11283.

[121] Park S, Daily JW, Zhang X, Jin HS, Lee HJ, Lee YH. Interactions with the MC4R rs17782313 variant, mental stress and energy intake and the risk of obesity in genome epidemiology study. Nutr Metab (Lond). 2016;**13**:38. DOI:10.1186/s12986-016-0096-8.

[122] González-Villalva A, Colín-Barenque L, Bizarro-Nevares P, Rojas-Lemus M, Rodríguez-Lara V, García-Pelaez I, et al. Pollution by metals: Is there a relationship in glycemic control? Environ Toxicol Pharmacol. 2016;**46**:337–343. http://doi.org/10.1016/j.etap.2016.06.023.

[123] Cappelletti R, Ceppi M, Claudatus J, Gennaro V. (2016). Health status of male steel workers at an electric arc furnace (EAF) in Trentino, Italy. J Occupl Med Toxicol (Lond). 2016;**11**:7. http://doi.org/10.1186/s12995-016-0095-8.

[124] Bahadar H, Mostafalou S, Abdollahi M. Growing burden of diabetes in Pakistan and the possible role of arsenic and pesticides. J Diabetes Metab Dis. 2014;**13**:117. **DOI:** 10.1186/s40200-014-0117-y.

[125] Bräuner EV, Nordsborg RB, Andersen ZJ, Tjønneland A, Loft S, Raaschou-Nielsen O. Long-term exposure to low-level arsenic in drinking water and diabetes incidence: A prospective study of the diet, cancer and health cohort. Environ Health Perspect. 2014;**122**:1059–1065. DOI:10.1289/ehp.1408198.

[126] Mendez MA, González-Horta C, Sánchez-Ramírez B, Ballinas-Casarrubias L, Hernández Cerón R, Viniegra Morales D, et al. Chronic exposure to arsenic and markers of cardiometabolic risk: A cross-sectional study in Chihuahua, Mexico. Environ Health Perspect. 2016;**124**:104–111.

[127] Moon KA, Guallar E, Umans JG, Devereux RB, Best LG, Francesconi KA, et al. Association between exposure to low to moderate arsenic levels and incident cardiovascular disease. A prospective cohort study. Ann Intern Med. 2013;**159**:649–659.

[128] Jia G, Aroor AR, Martinez-Lemus LA, Sowers JR. Mitochondrial functional impairment in response to environmental toxins in the cardiorenal metabolic syndrome. Arch Toxicol. 2015;**89**:147–153.

[129] Peng S, Liu L, Zhang X, Heinrich J, Zhang J, Schramm KW. et al. A nested case-control study indicating heavy metal residues in meconium associate with maternal gestational diabetes mellitus risk. Enviro Health. 2015;**14**:19. DOI: 10.1186/s12940-015-0004-0.

[130] Feseke SK, Ayotte P, Bouchard M, Levallois P. (Arsenic exposure and type 2 diabetes : Results from the 2007–2009 Canadian health measures survey. Health Promot Chronic Dis Prev Can. 2015;**35:**63–72.

[131] Rahman M, Tondel M, Chowdhury I, Axelson O. Relations between exposure to arsenic, skin lesions, and glucosuria. Occup Environ Med. 1999;**56**:277–281.

[132] Chihuahua ACS, Mendez MA, González-Horta C, Sánchez-Ramírez B, Ballinas-Casarrubias L, Cerón RH, et al. Chronic exposure to arsenic and markers of cardiometabolic risk. Environ Health Perspect. 2016;**104**:104–111.

[133] Peng Q, Harlow SD, Park SK. Urinary arsenic and insulin resistance in US adolescents. Int J Hyg Environ Health. 2015;**218**:407–413. doi:10.1016/j.ijheh.2015.03.006.

[134] Martin E, Gonzalez-Horta C, Rager J, Bailey KA, Sanchez-Ramirez B, Ballinas-Casarrubias L, et al. Metabolomic characteristics of arsenic-associated diabetes in a prospective cohort in Chihuahua, Mexico. Toxicol Sci. 2015;**144**:338–346. DOI: 10.1093/toxsci/kfu318.

[135] Kuo C-C, Howard BV, Umans JG, Gribble MO, Best LG, Francesconi K, et al. Arsenic exposure, arsenic metabolism, and incident diabetes in the strong heart study. Diabetes Care. 2015;**38**:620–627.

[136] Balakrishnan P, Vaidya D, Franceschini N, Voruganti VS, Gribble MO, Haack K, et al. Association of cardiometabolic genes with arsenic metabolism biomarkers in American Indian communities: The strong heart family study (SHFS). Environ Health Perspect. 2016 Jun 28. [Epub ahead of print]

[137] Wang ZX, Jiang CS, Liu L, Wang XH, Jin HJ, Wu Q et al. The role of Akt on arsenic trioxide suppression of 3T3-L1 preadipocyte differentiation. Cell Res. 2005;**15**:379–386.

[138] Divya SP, Pratheeshkumar P, Son YO, Roy RV, Hitron JA, Kim D, et al. Adipocytes and myotubes via oxidative stress-regulated mitochondrial sirt3-FOXO3a signaling pathway. Toxicol Sci. 2015;**146**:290–300. DOI:10.1093/toxsci/kfv089.

[139] Flora SJ. Arsenic-induced oxidative stress and its reversibility. Free Radic Biol Med. 2011;**51**:257–281.

[140] Liu S, Guo X, Wu B, Yu H, Zhang X, Li M. Arsenic induces diabetic effects through beta-cell dysfunction and increased gluconeogenesis in mice. Sci Rep.. 2014;**4**:6894.

[141] Simeonova PP, Hulderman T, Harki D, Luster MI. Arsenic exposure accelerates atherogenesis in apolipoprotein E-/- mice. Environ Health Perspect. 2003;**111**:1744–1748. DOI: 10.1289/ehp.6332.

[142] Lemaire M, Negro Silva LF, Lemarié CA, Bolt AM, Flores MOlina M, Krohn RM, et al. Arsenic exposure increases monocyte adhesion to the vascular endothelium, a pro-atherogenic mechanism. PLoS ONE. 2015;**10:**e0136592. DOI:10.1371/journal.pone.0136592.

[143] Kim KN, Park SJ, Choi B, Joo NS. Blood mercury and insulin resistance in nondiabetic Koreans (KNHANES 2008-2010). Yonsei Med J. 2015;**56**:944–950. DOI:10.3349/ymj.2015.56.4.944.

[144] He K, Xun P, Liu K, Morris S, Reis J, Guallar E. Mercury exposure in young adulthood and incidence of diabetes later in life. Diabetes Care. Jun 2013;**36**:1584–1589. **DOI:** 10.2337/dc12-184.

[145] Moon SS. Association of lead, mercury and cadmium with diabetes in the Korean population: The Korea National Health and Nutrition Examination Survey (KNHANES) 2009–2010. Diabet Med. 2013;**30**:e143–e148.

[146] Kobal AB, Prezelj M, Horvat M, Krsnik M, Gibicar D, Osredkar J. Glutathione level after long-term occupational elemental mercury exposure. Environ Res. 2008;**107**:115–123.

[147] Al-azzawie HF, Umran A, Hyader NH. Oxidative stress, antioxidant status and DNA damage in a mercury exposure workers. Br J Pharmacol Toxicol. 2013;**4**:80–88.

[148] Zeneli L, Sekovanić A, Ajvazi M. Kurti L, Daci N. Alterations in antioxidant defense system of workers chronically exposed to arsenic, cadmium and mercury from coal flying ash. Environ Geochem Health. 2016;**38**:65–72. DOI:10.1007/s10653-015-9683-2.

[149] Tinkov AA, Ajsuvakova OP, Skalnaya MG, Popova EV, Sinitskii AI, Nemereshina ON, et al. Mercury and metabolic syndrome: A review of experimental and clinical observations. BioMetals. 2015;**28**:231–254. DOI: 10.1007/s10534-015-9823-2.

[150] Ya WC, Chun FH, Keh ST, Rong SY, Cheng CY, Ching YY, et al. The role of phosphoinositide 3-kinase/Akt signaling in low-dose mercury-induced mouse pancreatic -cell dysfunction in vitro and in vivo. Diabetes. 2006;**55**:1614–1624. DOI: 10.2337/db06-0029.

[151] Barnes DM, Kircher EA. Effects of mercuric chloride on glucose transport in 3T3-L1 adipocytes. Toxicol In Vitro. 2005;**19**:207–214.

[152] Milnerowicz H, Sciskalska M, Dul M. Pro-inflammatory effects of metals in persons and animals exposed to tobacco smoke. J Trace Elem Med Biol. 2015;**29**:1–10. DOI: 10.1016/j.jtemb.2014.04.008.

[153] Tellez-Plaza M, Navas-Acien A, Caldwell KL, Menke A, Muntner P, Guallar E. Reduction in cadmium exposure in the United States population, 1988–2008: The contribution of declining smoking rates. Environ Health Perspect. 2012;**120**:204–209. DOI:10.1289/ehp.1104020.

[154] Will JC, Galuska DA, Ford ES, Mokdad A, Calle EE. Cigarette smoking and diabetes mellitus: Evidence of a positive association from a large prospective cohort study. Int J Epidemiol. 2001;**30**:540–546.

[155] Sargeant LA, Khaw KT, Bingham S, Day NE, Luben RN, Oakes S, et al. Cigarette smoking and glycaemia: The EPIC-Norfolk study. European prospective investigation into cancer. Int J Epidemiol. 2001;**30**:547–554.

[156] Adhami N, Starck SR, Flores C, Martins Green M. A health threat to bystanders living in the homes of smokers: How smoke toxins deposited on surfaces can cause insulin resistance. PLoS ONE. 2016;**11**:e0149510. DOI:10.1371/journal.pone.0149510.

[157] Afridi HI, Talpur FN, Kazi TG, Brabazon D. Effect of trace and toxic elements of different brands of cigarettes on the essential elemental status of Irish referent and diabetic mellitus consumers. Biol Trace Elem Res. 2015;**167**:209. DOI:10.1007/s12011-015-0308-5).

[158] Yang A, Cheng N, Pu H, Liu S, Dai M, Zheng T, Bai Y. Occupational metal exposures, smoking and risk of diabetes and prediabetes. Ocupp Med (Lond). July 14, 2016., pli. kqw078 [Epub ahead of print].

[159] Bizoń A, Antonowicz-Juchniewicz J, Milnerowicz M, Śliwińska-Mossoń M, Milnerowicz H. The effect of occupational exposure on pro/antioxidant balance in the blood of non-smoking and smoking smelters with diabetes. Environ Toxicol Pharmacol. 2016;**44**:99–106.

[160] Turyk M, Anderson H, Knobeloch L, Imm P, Persky V. Organochlorine exposure and incidence of diabetes in a cohort of great lakes sport fish consumers. Environ Health Perspect. 2009;**117**:1076–1082.

[161] Pesatori AC, Zocchetti C, Guercilena S, Consonni D, Turrini D, Bertazzi PA. Dioxin exposure and non-malignant health effects: A mortality study. Occup Environ Med. 1998;**55**:126–131.

[162] Vena J, Boffetta P, Becher H, Benn T, Bueno de Mesquita HB, Coggon D, et al. Exposure to dioxin and nonneoplastic mortality in the expanded IARC international cohort study of phenoxy herbicide and chlorophenol production workers and sprayers. Environ Health Perspect. 1998;**106**:645–653.

[163] Shapiro GD, Dodds L, Arbuckle TE, Ashley-Martin J, Fraser W, Fisher M, et al. Exposure to phthalates, bisphenol A and metals in pregnancy and the association with impaired glucose tolerance and gestational diabetes mellitus: The MIREC study. Environ Int. 2015;**83**:63–71. DOI: 10.1016/j.envint.2015.05.016.

[164] García de Aguinaga ML, Posada de la Paz M, Estirado de Cabo E, del Mar Plaza Cano M, Alvarez CS, Saro BB, et al. High prevalence of cardiovascular risk in patients with toxic oil syndrome: A comparative study using the general Spanish population. Eur J Intern Med. 2008;**19**:32–39.

[165] Chang JW, Chen HL, Su HJ, Lee CC. Abdominal obesity and insulin resistance in people exposed to moderate-to-high levels of dioxin. PLoS ONE. 2016;**11**:1–14.

[166] Imbeault P, Findlay CS, Robidoux MA, Haman F, Blais JM, Tremblay A, et al. Dysregulation of cytokine response in Canadian first nations communities: Is there an association with persistent organic pollutant levels? PLoS ONE. 2012;**7**:1–8. http://doi.org/10.1371/journal.pone.0039931.

[167] Fujiyosh PT, Michalek JE, Matsumura F. Molecular epidemiologic evidence for diabetogenic effects of dioxin exposure in U.S. air force veterans of the Vietnam war. Environ Health Perspect. 2006;**114:**1677–1683. http://doi.org/10.1289/ehp.9262.

[168] Chang J, Chen H, Su H, Liao P, Guo H, Lee C. Dioxin exposure and insulin resistance in taiwanese living near a highly contaminated area. Epidemiol. 2010;**21**:56–61.

[169] Qiu T, Chen M, Sun X, Cao J, Feng C, Li D, et al. Perfluorooctane sulfonate-induced insulin resistance is mediated by protein kinase B pathway. Biochem Biophys Res Commun. 2016;**477**:781–785.

[170] Seaton A, MacNee W, Donaldson K, Godden D. Particulate air-pollution and acute health-effects. Lancet. 1995;**345**:176–178.

[171] Eze IC, Hemkens LG, Bucher HC, Hoffmann B, Schindler C, Kunzli N, et al. Association between ambient air pollution and diabetes mellitus in Europe and North America: Systematic review and meta-analysis. Environ Health Perspect. 2015;**123**:381–389. DOI:10.1289/ehp.1307823.

[172] Chuang K, Yan Y, Cheng T. Effect of air pollution on blood pressure, blood lipids, and blood sugar: A population-based approach. J Occup Environ Med. 2010;**52**:258–262.

[173] Hansen AB, Ravnskjær L, Loft S, Andersen KK, Bräuner EV, Baastrup R, et al. Long-term exposure to fine particulate matter and incidence of diabetes in the Danish nurse cohort. Environ Int. 2016;**91**:243–50.

[174] Meo SA, Suraya F. Effect of environmental air pollution on cardiovascular diseases. Eur Rev Med Pharmacol Sci. 2015;**19**:4890–4897.

[175] Zheng Z, Xu X, Zhang X, Wang A, Zhang C, Huttermann M, et al. Exposure to ambient particulate matter induces a NASH-like phenotype and impairs hepatic glucose metabolism in an animal model. J Hepatol. 2013;**58**:148–154.

[176] Dennekamp M, Mehenni O, Cherrie JW, Seaton A. Exposure to ultrafine particles and PM2. 5 in different micro-environments. Ann Occup Hyg. 2002;**46**:412–414.

[177] European Environmental Agency. Emissions of primary PM2.5 and PM10 particulate matter, 2014. http://www.eea.europa.eu/data-and-maps/indicators/emissions-of-primary-particles-and-5/assessment-3. [Accessed 2016:08:15]

[178] Shi L, Zanobetti A, Kloog I, Coull BA, Koutrakis P, Melly SJ, et al. Low-concentration PM2.5 and mortality: Estimating acute and chronic effects in a population-based study. Environ Health Perspect. 2016;**124**:46–52. http://dx.doi.org/10.1289/ehp.1409111.

[179] Pope CA, Turner MC, Burnett RT, Jerrett M, Gapstur SM, Diver WR, et al. Relationships between fine particulate air pollution, cardiometabolic disorders, and cardiovascular mortality. Circ Res. 2015;**116**:108–115. http://doi.org/10.1161/CIRCRESAHA.116.305060.

[180] Choi YH, Kim JH, Hong YC. Sex-dependent and body weight-dependent associations between environmental PAHs exposure and insulin resistance: Korean urban elderly panel. J Epidemiol Commun Health. 2015;**69**:625–631. DOI:10.1136/jech-2014-204801.

[181] Kong S, Ding X, Bai Z, et al. A seasonal study of polycyclic aromatic hydrocarbons in PM(2.5) and PM(2.5–10) in five typical cities of Liaoning Province, China. J Hazard Mater. 2010;**183**:70–80.

[182] Saldarriaga-Noreña H, López-Márquez R, Murillo-Tovar M, Hernández-Mena L, Ospina-Noreña E, Sánchez-Salinas E, et al. Analysis of PAHs associated with particulate matter PM2.5 in two places at the city of Cuernavaca, Morelos, México. Atmosphere. 2015;**6**:1259–1270.

[183] Wei Y, Zhang J, Li Z, Gow A, Chung KF, Hu M, et al. Chronic exposure to air pollution particles increases the risk of obesity and metabolic syndrome: Findings from a natural experiment in Beijing. FASEB J. 2016;**30**:2115–2122.

[184] Khatri M, Bello D, Pal AK, Cohen JM, Woskie S, Gassert T, et al. Evaluation of cytotoxic, genotoxic and inflammatory responses of nanoparticles from photocopiers in three human cell lines. Part Fibre Toxicol. 2013;**10**:1–22. http://doi.org/10.1186/1743-8977-10-42.

[185] Brocato J, Hernandez M, Laulicht F, Sun H, Shamy M, Mansour A, et al. In vivo exposures to particulate matter collected from Saudi Arabia or nickel chloride display similar dysregulation of metabolic syndrome genes. J Toxicol Environ Health Part A. 2015;**78**:1421–1436. DOI:10.1080/15287394.2015.1095689.

[186] Qi L, Cornelis MC, Kraft P, Jensen M, van Dam RM, Sun Q, et al. Genetic variants in ABO blood group region, plasma soluble E-selectin levels and risk of type 2 diabetes. Hum Mol Gen. 2010;**19**:1856–1862.

[187] Suadicani P, Hein HO, Gyntelberg F. Airborne occupational exposure, ABO phenotype and risk of ischaemic heart disease in the copenhagen male study. J Cardiovasc Risk. 2002;**9**:191–198.

[188] Wu O, Bayoumi N, Vickers MA, Clark P. ABO(H) blood groups and vascular disease: A systematic review and meta-analysis. J Thromb Haemos. 2008;**6**:62–69.

[189] Kristen E, Cosselman E, Krishnan R. Blood pressure response to controlled diesel exhaust exposure in human subjects. Hypertension. 2012;**59**:943–948.

[190] Devlin RB, Smith CB, Schmitt MT, Rappold AG, Hinderliter A, Graff D, et al. Controlled exposure of humans with metabolic syndrome to concentrated ultrafine ambient particulate matter causes cardiovascular effects. Toxicol Sci. 2014;140:61–72.

[191] Baja ES, Schwartz JD, Wellenius GA, Coull BA, Zanobetti A, Vokonas PS, et al. Traffic-related air pollution and QT interval: Modification by diabetes, obesity, and oxidative stress gene polymorphisms in the normative aging study. Environ Health Perspect. 2010;**118**:840–846.

Chapter 7

Epidemiology of Needlestick and Sharps Injuries in Veterinary Medicine

Ana Cláudia Coelho

Additional information is available at the end of the chapter

http://dx.doi.org/10.5772/66110

Abstract

Needlestick injuries (NSIs) are a serious concern for veterinary practitioners as well as other healthcare personnel. During practice, veterinarians are exposed to various risky situations in which NSI and sharps injuries seem to be a common occupational hazard. Studies on prevalence and occurrence of NSI in veterinary medicine are scarce and probably underreported. One important consequence is the physical trauma. However, other factors related to their economic or psychiatric impact should also be considered. The studies available about NSI in veterinarians reported different prevalence, ranging from 1 % to 86.7 %, although their comparison is difficult since prevalence is calculated from different data sources. Various risk factors of NSI (such as years as veterinarians, number of work hours, poor quality of restraint of animals, poor needle handling practices, among others) have been described. However, information regarding risk factors in veterinary medicine is scarce. In order to understand the epidemiology of NSI in veterinarians, a review of the literature published in the last four decades (1980–2016) is presented. Thus, the current chapter will address several characteristics of NSI in veterinary medicine as occurrence, prevalence and incidence risk factors, consequences and preventive measures.

Keywords: needlestick, sharps, injury, epidemiology, risk factor, prevalence

1. Introduction

Occupational health problems in veterinary medicine are very frequent, and veterinarians are considered to be members of a high-risk group for occupational hazards [1].

A needlestick injury (NSI) can be defined as an inadvertent (accidental) penetrating wound from a needle that may result in exposure to the blood or other body fluids. A sharps injury includes needles or other sharp objects, such as scalpels, lancets, razor blade, scissors, nose tongs for cattle, halters, calf pulling equipment and metal cattle chutes [2, 3]. These types of injuries are considered a major occupational health problem and of serious concern for veterinarians and other healthcare workers [4]. NSI injuries usually occur during activities such as taking blood and body fluid specimens and processing, needle disposal, waste collection and transferring blood from a syringe into another vessel [5].

Awareness of the transmissibility of bloodborne infectious agents in human medicine, including human immunodeficiency virus (HIV), hepatitis B virus and hepatitis C virus, has led to the identification of percutaneous sharps injury resulting in exposure to bloodborne pathogens as an important occupational health risk for people employed in the healthcare industry [6]. It is estimated that more than 2 million healthcare workers experience an NSI or sharps injury with a contaminated sharp instrument every year [7]. Injuries associated with NSI are associated with the potential exposure to infectious agents and syringe contents [8]. Injuries due to contact with contaminated needles may also have serious physical and psychological consequences [9].

There has been less concern regarding NSI and sharps injuries in veterinary practice, and only a few epidemiological studies have been conducted in this area. On the other hand, methodological aspects are not comparable with different design approaches [8]. The analysis of NSI is essential to identifying areas of improvement.

In order to understand the epidemiology of NSI in veterinarians over the last four decades, in this chapter we review the literature, focusing on the epidemiology of NSI and sharps injuries in veterinary practice.

2. Epidemiology of needlestick and sharps injuries

Epidemiologic data on NSI and sharps injuries are essential for targeting and assessing interventions [10]. However, a few studies have looked at the epidemiology of NSI and sharps injuries in veterinary practice. Despite significant effort for reduction, NSI and sharps injuries continue to pose a significant risk in human medicine [11], and a similar risk occurs in veterinary medicine. Reports carried out in small and large-animal practice show a large variability in the prevalence and incidence.

2.1. Biological risk and bloodborne infections

Bloodborne infections are recognized for a long time, and they are the main risk to the health of workers exposed to blood and other biological materials. However, it was only after the discovery of HIV that occupational injuries with potentially contaminated biological material were treated as a public health problem [12]. Infection control guidelines in human medicine put emphasis on protection against bloodborne pathogens. In veterinary medicine few serious zoonotic infections are currently considered to be bloodborne which reduces concern in the veterinary community [13]. HIV and hepatitis viruses are of potential concerns in human medicine but are absent in veterinary medicine [2]. However, veterinarians, by maintaining direct contact with animals, are often exposed to biological agents found in the blood and body fluids. In veterinary hospitals and clinics, occupational risk by biological agents is universally distributed; the risks are proportional to the amount of contacts with patients and blood, secretions and other body fluids [14, 15].

The risk of transmission of infectious agents after the injury with biological material depends on several factors such as host susceptibility and resistance, virulence of the agent, the route of exposure and amount of the infectious agent. As such, the greater the manipulation of sharp objects, blood and other body fluids, the greater the exposure and risk of acquiring infectious diseases [16].

The few recognized bloodborne pathogens that can be transmitted between animals and humans probably are the reason for the less concern regarding NSI and sharps in veterinary practice [17]. The most important bloodborne pathogens in veterinary work are *Staphylococcus* spp.; *Pseudomonas* spp. (inoculated from the animal skin); pathogens from fine-needle aspirates *Blastomyces*, *Pasteurella* spp., *Staphylococcus* spp. and *Streptococcus* spp. (from fine-needle aspirates); certain arboviruses or modified live vaccines [2]. *Bartonella* spp. appear to be a zoonotic pathogen [18]. There are two reports in the literature about accidental needlestick transmission of *Bartonella* to veterinarians. One study reported a suspected needlestick transmission of *Bartonella vinsonii* subspecies *berkhofii* to a veterinarian [19], and another reports a veterinarian with *Bartonella henselae* after a needle puncture [20]. Disease associated with NSI occurrence was addressed in one study [21] where injury during *Brucella* vaccine administration was found to be a risk factor for occupational brucellosis. More reports in veterinary practice that NSI and sharps injuries resulted in a zoonotic disease are of a 26-year-old veterinary technician who became infected with B virus following a needlestick injury [22] and blastomycosis developed in a veterinarian after an NSI following a fine-needle aspiration [23]. No cases of other biological agents have been reported after NSI in small or large-animal practice in the literature review. Exotic zoonotic pathogens are hypothetically transmitted through contact with the blood, and the risk of emerging bloodborne pathogens should be seriously considered by all veterinary practitioners [13, 24, 25].

2.2. Prevalence and incidence of NSI and sharps injuries in veterinarians

Approximately 1.2 million occupational NSI and sharps injuries occur in the European Union (EU) each year [26]. It is very difficult to conduct studies in prevalence or incidence (per time per person) or exposure rate of NSI and sharps injuries in veterinary medicine, but these studies are important to determine the risk factors associated with occurrence of injuries [1]. Another problem is that the rate of underreporting of NSI and sharps events is very high [27], and the quality of available data is variable [2].

The studies available about NSI and sharps injuries in veterinary medicine reported different prevalence, ranging from 1% to 86.7%. However, their comparison is difficult since prevalence is calculated from different data sources. In one study conducted in veterinarians in Wisconsin, the incidence of NSI and sharps exposures to Johne's bacterin during vaccination against paratuberculosis was 5.5/100 person-years [28]. A survey conducted in female veterinarians reported 63.9% of one or more needlesticks after graduation from veterinary college. The incidence of NSI was 9.3/100 person-years [27]. The prevalence of NSI and sharps injuries in American zoo veterinarians was 86.7% [29]. The overall exposure rate reported by Australian veterinarians was 75.3%, but those reported suffering from at least one contaminated NSI in the previous year were 58.9% [30]. A survey of veterinary technicians reported that 93% had at least one NSI over the course of their career and 74% had experienced a needlestick injury during the previous 12 months of the study [8]. In another study, veterinarians that reported at least one unintentional NSI were 74.2 in the previous year of the study [31]. A survey of veterinarians from Uganda reported a NSI prevalence of 15.0% [32]. In a study performed in Portugal, 78.5% of veterinarians enrolled in the study reported having had at least one NSI during their careers [33]. The prevalence of NSI in Japanese veterinarians during containment measures of foot-and-mouth diseases was 1%, and NSI accounted for 18% of all reported injuries in all veterinarians [34].

Rates observed in veterinary medicine are variable, but in some epidemiological studies, values are much higher than the prevalence rates described for human medicine [35, 36].

2.2.1. Reporting

Accurate reporting of NSI and sharps injuries is essential, to ensure that incidents are appropriately managed [37]. An accident with a needlestick or a sharp should be reported immediately to the supervisor, which supports the workers in administrative and legal terms if they develop a disease resulting from an accident [38]. Surveillance in NSI and sharps injuries should be activated in every healthcare setting to monitor injuries and contaminations and identify the need for corrective interventions [39]. It is difficult to provide accurate statistics on the incidence of NSI or sharps injuries because even in developed countries in human medicine, all cases are not reported [40]. Reasons for underreporting in human medicine include the lack of necessity of reporting with a presumption that the risk of bloodborne pathogens transmission is low and lack of knowledge of systems [37]. There is no single reporting system for injuries or disease in veterinarians, and reported cases of NSI and sharps may greatly underestimate the real number of occurrences [1]. Recall bias and deterioration of memory with the passage of time are other problems associated with the rates of prevalence reported. The true incidence may be underreported owing to the incapacity of busy professionals to remember and write down the details [27].

2.3. Risk factors of NSI and sharps injuries in veterinarians

Identification of risk factors of NSI and sharps has been reported in few epidemiological studies involving veterinarians. As a consequence, little information is available concerning the risk factors for NSI and sharps injuries. Some risk factors associated with NSI and sharps have been referred to in the literature as presented in **Table 1**.

Factors described	Reference
Poor quality of restraint	[30]
Inadequate access to sharps containers	[49]
Poor needle handling practices by veterinarians	[13, 29, 52]
Female gender	[30]
Working in small-animal practice	[30]
Working with large animals	[46, 47]
Veterinarians working with dogs	[33]
Less experienced veterinarians (years as veterinarian)	[30, 41]
Veterinarians with more than 10 years of practice	[33]
Seeing excessive numbers of patients per week	[30]
Working longer than normal hours per week	[30]
Number of patients treated per week	[30]
Household bovines and sheep during childhood	[33]

Table 1. Factors described as contributing to the risk of NSI and sharps injuries in veterinarians.

2.3.1. Demographic and workplace items

The influence of sex in the prevalence of NSI and sharps injuries has been studied. Female veterinarians presented higher odds of injury than male veterinarians [30].

Years as veterinarians have been described as risk factors. Less experienced veterinarians [30, 41] were more likely to report injuries, which is consistent with that observed in human medicine where the probability of injuries by sharp devices among new personnel or healthcare students is superior when compared to healthcare workers with more years of experience [42]. Clinical experience may have provided expertise and techniques for handling needles and sharp devices, reducing the risk of occupational injuries [43]. However, other studies of veterinarians are contradictory and showed that the proportion of veterinarians who experienced NSI increases with years of practice [33]. This is probably because younger veterinarians may apply the knowledge in infection control acquired in the university and put safety procedures into practice, while practitioners with experience have familiarity with needles and sharp instruments and pay less attention to risks and have lower compliance with biosafety measures [44].

In human medicine, increased risk of injury incidents is positively associated with time constraints and rushing to complete procedures, nervousness, tiredness and loss of concentration. The predisposition to increased risk was also noted to be associated with high workload, working hastily, a crowded work environment, times when personnel are fatigued, do not have a patient's collaboration or when the medical team was not fully present [12, 45]. In veterinary medicine there are no studies that associate time constraints with NSI and sharps injuries, but these injuries were directly connected with working longer hours per week and number of patients treated per week [30].

Household association with cattle during childhood has previously been identified as a risk factor. Interaction with animals in infancy could lead to a sense of security in the handling of animals. This can give rise to overconfidence and generate negligent safety behaviours [33].

2.3.2. Type of practice

There is no consensus about the link between the type of practice and the occurrence of accidents that cause injuries. Working in small or mixed animal practice was associated with a significantly higher exposure rate for contaminated NSI in one study [30]. In another study, working with dogs was a risk factor for NSI occurrence, probably, because these animals are extremely mobile animals that are seldom fully restrained during the course of veterinary care and they are often treated with parenteral drugs [33]. According to another study, people working with large animals were more likely to report injuries [46] suggesting that the treating of large animals is more hazardous than the treating of small animals [47]. Animal handling and environmental problems probably influence the occurrence, such as working in semi-dark settings, confinement in closed spaces and high animal densities. However, in a Minnesota and Wisconsin survey, the type of practice did not affect the exposure rate to NSI and sharps injuries, although large animals caused more severe injuries—which is not surprising [48].

Veterinary occupational injury can increase with prior injuries, participation in sports, current smoking and six or fewer hours of sleep [49]. Neglected management of occupational health and a failure to comply with simple proactive measures are risk factors for NSI events [38].

2.3.3. Working conditions

2.3.3.1. Poor quality of restraint

Poor quality of restraint caused by lack of adequate personnel or inadequate assistance with restraint of animals is considered as a risk factor for an NSI event. Animals are far less obedient than human patients, and movement of the animal at the time of needle puncture is more common if the animal patient is not well restrained [30]. It is probable that the large-animal veterinarians may experience a lower rate of needlestick puncture wounds because they are more likely to restrain their large-animal patients compared to the small-animal clinicians [27, 50]. Poor restraint can compromise not only the person inoculating but also other healthcare workers and animal owners who may be helping [2].

2.3.3.2. Inadequate access to sharps containers

No sharps boxes present increased rates to injuries [49]. Sharps should be disposed of immediately after use directly into a container (i.e. not left bare on any surface) [17]. These sharps containers need to be in close proximity so that the staff can place the sharps into the appropriate containers immediately after use [51]. Pocketing of needles poses a risk for NSI to other staff whilst doing unrelated tasks such as in the laundry [2, 8]. Poor needle handling practices, such as not firmly recapping needles or not bothering to recap them at all, is an important risk factor [13, 29, 52]. But it is likely that this behaviour is improving. In a study performed in small-animal veterinary practices, 89% of practices dispose of sharps directly, rather than recapping them, though a few even indicated a preference for careful recapping [53]. If an appropriate disposal container is not in proximity and available, recapping should be done using some procedures and techniques such as the one-handed 'scoop' or using a device to handle the needle cap [54, 55].

2.4. Procedures associated with NSI and sharps injuries

Needlestick and sharps injuries most often occur before disposal of a needle or sharp device, during the use of a sharp device, after a procedure and after improper disposal (leaving needles in a laboratory coat with subsequent needlestick injury to laundry personnel) [2]. There are many possible mechanisms of injury. Some circumstances associated with NSI and sharps have been referred to in the literature.

Regarding the circumstances in which incidents of NSI and injuries by sharp instruments occurred, it is possible to observe that injury can occur during vaccination and other procedures, in which veterinarians are frequently accidentally 'self-inoculated' or suffer other self-inflicted wounds [20, 31, 32, 56, 57]!

2.4.1. Vaccine administration

Concerning veterinary activity at time of injury, it has been demonstrated that vaccine administration is an activity that accounts for a lot of accidents. Within the studies reviewed which evaluated substances involved, two studies refer to injury during *Brucella* vaccine administration, RB51 [58] and S19 strain [21]. Vaccine administration in chickens against infectious bursal disease (Gumboro) and Newcastle disease was the single most important cause of self-inoculations in practitioners from Uganda [32]. A study of 1347 NSI involving vaccines demonstrated that one-third of the vaccine-related sticks involved rabies and about 11% involved distemper vaccines [27]. Self-inoculation with the vaccine against *Leptospira* was reported by 7.5% of US veterinarians [31]. Accidental injection of an inoculation against (*Mycobacterium avium* subspecies *paratuberculosis*) (bovine Johne's disease) was reported in two studies [28, 56]. Accidental exposures (due to NSI) to vaccines against diseases such as West Nile virus, *Giardia* and *Leptospira* spp. [46] and to live equine vaccination against Equilis StrepE [57] have also been reported in the literature. Erysipelas vaccine and other vaccines was the most commonly cited agent exposure reported by US swine veterinarians in a study of occupational hazards [52].

2.4.2. Animals involved

Procedures involving large animals were reported in more studies [27, 29, 30, 32, 34, 46, 47, 52, 58, 59] of epidemiology occurrence than small animals. Zoo animals were also involved in accidents [29].

2.4.3. Poor infection control practices

Recapping needles is another activity that increased the risk of percutaneous injury. There was an association between recapping needles in small-animal practice and in large-animal practice and NSI and sharps injuries [13, 52]. A study of zoo veterinarians reported that 86.0 % of NSIs involved recapping needles [29]. Uncapping of needles by the mouth can be a relatively common but risky form of behaviour [17]. Another practice of risk for parenteral exposures, especially in large-animal medicine, is the reuse of needles and syringes [13]. Standard precautions for human medicine guidelines recommend never recapping needles [51].

2.4.4. Injury location

There was little information about the most affected anatomical parts of the body injured. There was no agreement between studies with respect to the most frequent site of the involvement. In one study, a veterinarian experienced an NSI in the right index finger [19]; in other two studies, veterinarians experienced an NSI into the right thumb [20, 56]; and in another study, seven veterinarians were said to have experienced an NSI in the upper limb [32].

2.5. Consequences and side effects of NSI and sharps injuries

The consequences of occupational exposure to the blood and other body fluids are not only related to infections but also the psychological trauma, anxiety, relationships and prophylactic drugs [60]. NSI and sharps injuries can produce physical trauma, but it is unlikely that they cause severe injuries alone. Physical trauma such as severe laceration can be significant, especially from large-bore needles, and can result from animal movement during injection or blood collection [2]. Every needlestick and sharps injury carries a risk of trauma or inoculation of harmful substances. While the physical trauma caused by needle or a sharp in the body may often be minor, introduction of hazardous compounds such as chemical or biological contaminants has been associated with severe sequelae, including serious infections and damage to tissue [27]. Side effects of NSI and sharps injuries following accidental exposures were normally characterized as mild or severe and local or systemic. Serious adverse effects, while uncommon, do occur [2]. Local adverse events are characterized by one or more of the following symptoms: pain, erythema, local swelling and superficial abscess [27, 52, 58]. These were frequently reported after an injury. Systemic adverse effects experienced after NSI or sharps injuries included myalgia, fever, arthralgia, headaches, fatigue, sweats, severe allergic reaction, chills, lacerations, psychedelic experience, diarrhoea, vomiting or granuloma [27, 56, 58, 59]. Severe reactions included severe local inflammation, abscess formation, localized necrosis, local nerve damage, disease, severe allergic reaction and miscarriage [2, 23].

Veterinarians experiencing adverse reactions are more likely to report having had a NSI than others [2].

Psychological and psychiatric consequences of NSIs are not yet quantified in veterinary medicine. In human medicine occupational blood exposure can lead to posttraumatic stress, anxiety and depression and is a major contributing factor of time loss from work [60–62].

2.5.1. *Syringes and needle content and side effects*

Although many types of sharps injure veterinarians, the most common causative devices associated with a higher rate of injury were syringes and needles [19, 30, 47], needle biopsy [56], scalpel blades [63] and ampoule/vial [30]. Although some accidents occurred with empty or clear needles [27, 52], NSI injuries may involve the risk of self-injecting drugs or other hazardous substances, which can result in mild or severe allergic reactions or other more severe consequences [2, 27].

Agents producing a side effect most often include anthelmintics [27, 52], euthanasia agents and anaesthetics and steroids [27], immobilizing agents [29], hormones, vaccines [20, 21, 28, 29, 31, 32, 46, 56] and antibiotics [29, 30, 32]. Mineral oil adjuvants of veterinary vaccines can produce a chronic granulomatous reaction with sterile abscess formation [64]. Accidental needlestick injuries and conjunctival or open wound exposures of humans involving the RB51 vaccine were associated with both local and systemic adverse events in the United States [58].

Occupational NSI and sharps injuries may also represent a serious human reproductive health hazard, notably the unintentional injection of dinoprost tromethamine, a prostaglandin compound leading to miscarriage in a previous study [27].

In some cases NSI and sharps injuries require medical treatment with hospital admission, in which case medical attention [20, 29, 34, 56, 57] and sometimes surgical intervention are needed [56, 58]. In a study examining zoo veterinarians, 6.5% of veterinarians required medical care after a NSI event [29]. The demand for medical treatment occurs in cases of adverse reactions to injected harmful substances and severe trauma. Self-treatment of injuries was common [48].

2.6. Prevention of needlestick and sharps injuries

Needlestick and sharps injuries are a serious problem in veterinary medicine, but it is often preventable. In human medicine almost 83% of needlestick injuries can be prevented [65]. However, preventive efforts can reduce the risk of exposures, but not eliminate them [66].

In human medicine, time and considerable economic resources have been expended to reduce the incidence of NSI associated with bloodborne agents. Some countries and governments have invested in the need to introduce safety devices, educating healthcare workers on the safe handling and disposal of sharp devices and developing strategies to prevent them [67, 68]. Aggressive educational campaigns concerning NSI prevention are lacking in veterinary medicine. Probably, the factor associated with the lack of this approach is a poorly developed culture of concern about biosafety in veterinary medicine, and only a few bloodborne zoonotic pathogens are recognized in clinically normal animals [2, 8].

2.6.1. *Safe practices*

Adherence and compliance with the universal precaution recommendations proposed by the Centers for Disease Control and Prevention (CDC) are important factors for the prevention of NSI [69]. Recommended prevention strategies include educational programmes, avoidance of recapping, better needle disposal systems and careful handling and disposal of sharp devices [12]. Infection control and workplace safety include safe handling of sharps [17]. Avoid recapping needles or use a 'one-handed scooping technique' to recap is a simple infection control procedure which may substantially reduce this form of occupational injury [46]. The CDC recommended the use of gloves and gowns during patient contact that requires handling of blood [69]. Although this equipment does provide a physical barrier to shield the skin and mucous membranes from contact with blood, most protective equipment is easily penetrated by needles [10]. Universal precautions recommended that persons that manipulate needles and other sharps wear gloves and have eye protection and reduce the risk of exposure to needlestick [58]. In previous studies, wearing two pairs of gloves seems to protect because when the outer glove is perforated, the inner glove can protect the hand [70, 71]. Previous studies suggest a low compliance with personal protective equipment in healthcare workers [39]. Discomfort, reduction of agility and decreased sensation of touch were reported to outweigh the benefits afforded by double gloving [72]. The use of personal protective equipment could be affected by availability [51].

The risk of NSI can also be reduced by the use of safety medical devices, which are becoming more commonly used in human healthcare [73] incorporating safety-engineered protection mechanisms (safety-engineered devices, e.g. retractable needles, fixed-needle safety syringes). These modern safety devices minimize the risk and impact of NSI injuries [39, 45]. However some healthcare workers refuse to use such devices [4]. Cost is an obvious concern with needle safety devices, particularly when the benefits are difficult to quantify [2]. And in veterinary medicine, the cost of using the safety-engineered devices can be unaffordable, and cost may be a limiting factor for the use of this kind of device [52].

Risk management prevention is necessary to reduce the likelihood of NSI and sharps injuries [2]. The implementation of legislation into the field can help identify and reduce future risk of these injuries [67, 74]. In Europe, legislation to improve the safety and health of personnel has been in place since 1989 and was published to protect healthcare workers and requires an integrated approach [74]. However, in some countries, this legislation is not adapted to veterinary medicine. Current guidelines to reduce NSI and sharps injuries in veterinary practice are not based on veterinary data, but are modified from studies in human medicine [30]. On the other hand, it is very important that veterinarians understand the reasons to comply with safe procedures, which include good needle and sharp handling practices and correct disposal by veterinarians [39]. Veterinarians should be familiar with the recommendations of the CDC guidelines on universal precautions [51]. Work-practice controls are important in preventing exposures to blood and hazardous substances and include verbal statements when passing sharps, avoiding hand-to-hand passage of sharp instruments [75].

2.6.2. Cost-benefit effectiveness of prevention

Costs are harder to quantify. They include the direct costs associated with the initial and follow-up treatment when necessary [76], the emotional cost associated with fear and anxiety associated with the possible consequences of the injury, direct and indirect costs related with lost productivity and cost of any associated legal action [10, 11]. Occupational hazards in the work of American veterinarians resulted in an estimated US$ 4 million in losses [77], and the costs of injuries related to sharp contaminated instruments in the USA have been estimated to be around 118–591 million dollars in 2010 [78]. These costs associated with NSI can be reduced and healthcare protected with investment in safety-engineered sharp devices [79].

2.6.3. *Education and training*

Education plays an important role in decreasing NSI rates as it decreased recapping, unnecessary needle manipulation and improper disposal of used devices [39]. Training should always be provided for new employees and periodically for veterinary clinical personnel, as well as for supporting staff [7]. Educational interventions including videotaping and performance feedback proved effective in the short term; however, long-term adherence was not observed [80]. To encourage constant compliance with good safety standards, educational sessions incorporating regular teaching, practical classes and reminders in the form of posters could be used [51]. Education and training need to be encouraged in older workers who receive less training and have more limited access to new technologies than younger workers [39, 81]. Adequate staffing and personnel training in proper animal restraint are also important, as poor restraint is an important risk factor [50]. To prevent injuries with aggressive animals, it is important to handle those animals with care and to make proper use of restraining devices and protective equipment [47].

3. Conclusions

This chapter describes the epidemiology of NSI and sharps injuries in veterinary medicine and emphasizes the importance of compliance with international standards of infection control practices, of training and of the education of veterinarians. It emphasizes the need for reporting and prevention of NSI and sharps injuries. Increasing awareness of hazards and how to avoid them and establishing better work environments are also crucial. Education regarding the use of personal protective equipment and the importance of reporting accidents should be promoted. There is a need to assess accurately the risk of NSI and sharps hazards in veterinary practice in order to develop effective measures for reducing related incidents. More epidemiological studies in this field are needed to study risk factors, to determine knowledge, attitudes and practices. It is also essential to put a cost-effective and efficient injury and control programme into place.

Acknowledgements

The work was supported by the strategic research project PEst-OE/AGR/UI0772/2014 financed by the Foundation for Science and Technology [FCT] and to Prof. D.A. Davis that fully revised English text.

Author details

Ana Cláudia Coelho

Address all correspondence to: accoelho@utad.pt

Animal and Veterinary Research Centre (CECAV), University of Trás-os-Montes and Alto Douro, Portugal

References

[1] Jeyaretnam J, Jones H. Physical, chemical and biological hazards in veterinary practice. Australian Veterinary Journal. 2000;78:751–758.

[2] Weese JS, Jack DC. Needlestick injuries in veterinary medicine. The Canadian Veterinary Journal. 2008;49(8):780.

[3] Bonini S, Buonacucina A, Selis L, Peli A, Mutti A, Corradi M. Occupational hazards in veterinarians: an updating. Journal of Veterinary Science & Technology. 2016;7:317. DOI: 10.4172/2157-7579.1000317

[4] Higginson R, Parry A. Needlestick injuries and safety syringes: a review of the literature. British Journal of Nursing. 2013;22(8):S4,S6–8. DOI: 10.12968/bjon.2013.22.Sup5.S4.

[5] Gabriel J. Reducing needlestick and sharps injuries among healthcare workers. Nursing Standard. 2009;23:41–44. DOI: 10.7748/ns2009.02.23.22.41.c6777.

[6] DeGirolamo KM, Courtemanche DJ, Hill WD, Kennedy A, Skarsgard ED. Use of safety scalpels and other safety practices to reduce sharps injury in the operating room: what is the evidence? Canadian Journal of Surgery. 2013;56:263–269. DOI: 10.1503/cjs.003812.

[7] Goniewicz M, Włoszczak-Szubzda A, Niemcewicz M, Witt M, Marciniak-Niemcewicz A, Jarosz MJ. Injuries caused by sharp instruments among healthcare workers – international and Polish perspectives. Annals of Agricultural and Environmental Medicine: AAEM. 2012;19:523–527.

[8] Weese JS, Faires M. A survey of needle handling practices and needlestick injuries in veterinary technicians. Canadian Veterinary Journal. 2009;50(12):1278–1282.

[9] Lee JM, Botteman MF, Xanthakos N, Nicklasson L. Needlestick injuries in the United States. Epidemiologic, economic, and quality of life issues. AAOHN Journal. 2005;53(3):117–133.

[10] Centers for Disease Control and Prevention. Sharps injury prevention workbook. 2015. 162 p. http://www.cdc.gov/sharpssafety/pdf/workbookcomplete.pdf. Accessed 17.08.16.

[11] Mannocci A, De Carli G, Di Bari V, Saulle R, Unim B, Nicolotti N, Carbonari L, Puro V, La Torre G. How much do needlestick injuries cost? A systematic review of the economic evaluations of needlestick and sharps injuries among healthcare personnel. Infection Control and Hospital Epidemiology. 2016;37(6):635–646. DOI: 10.1017/ice.2016.48.

[12] Centers for Disease Control and Prevention—CDC. Recommendations for prevention of HIV transmission in health care settings. Morbidity and Mortality Weekly Report. 1987;36(Suppl 2):3–18.

[13] Wright JG, Jung S, Holman RC, Marano NN, McQuiston JH. Infection control practices and zoonotic disease risks among veterinarians in the United States. Journal of the American Veterinary Medical Association. 2008;232(12):1863–1872. DOI: 10.2460/javma.232.12.1863.

[14] Davanzo E, Frasson C, Morandin M, Trevisan A. Occupational blood and body fluid exposure of university health care workers. American Journal of Infection Control. 2008;36:753–756. DOI: 10.1016/j.ajic.2008.04.254.

[15] MacCannell T, Laramie AK, Gomaa A, Perz JF. Occupational exposure of health care personnel to hepatitis B and hepatitis C: prevention and surveillance strategies. Clinical Liver Disease. 2010;14:23–36. DOI: 10.1016/j.cld.2009.11.001.

[16] Wicker S, Cinatl J, Berger A, Doerr HW, Gottschalk R, Rabenau HF. Determination of risk of infection with blood-borne pathogens following a needlestick injury in Hospital Workers. Annals of Occupational Hygiene. 2008;52:615–622.

[17] Anderson MEC, Weese JS. Video observation of sharps handling and infection control practices during routine companion animal appointments. BMC Veterinary Research. 2015;11:185. DOI: 10.1186/s12917-015-0503-9.

[18] Chomel BB, Boulouis HJ, Maruyama S, Breitschwerdt EB. *Bartonella* spp. in pets and effect on human health. Emerging Infectious Diseases. 2006;12:389–394. DOI: 10.3201/eid1205.050931

[19] Oliveira AM, Maggi RG, Woods CW, Breitschwerdt EB. Suspected needle stick transmission of *Bartonella vinsonii* subspecies *berkhoffii* to a veterinarian. Journal of Veterinary Internal Medicine. 2010;24:1229–1232. DOI: 10.1111/j.1939-1676.2010.0563.x.

[20] Lin JW, Chen CM, Chang CC. Unknown fever and back pain caused by *Bartonella henselae* in a veterinarian after a needle puncture: a case report and literature review. Vector-Borne and Zoonotic Diseases. 2011;11(5):589–591.

[21] Kutlu M, Ergonul O, Sayin-Kutlu S, Guven T, Ustun C, Alp-Cavus S, Ozturk SB, Acicbe O, Akalin S, Tekin R, Tekin-Koruk S, Demiroglu YZ, Keskiner R, Gönen I, Sapmaz-

Karabag S, Bosnak V, Kazak E. Risk factors for occupational brucellosis among veterinary personnel in Turkey. Preventive Veterinary Medicine. 2014;117(1):52–58.

[22] Artenstein AW, Hicks CB, Goodwin BS, Hilliard JK. Human infection with B virus following a needlestick injury. Reviews of Infectious Diseases. 1991;13:288–291.

[23] Ramsey D. Blastomycosis in a veterinarian. Journal of the American Veterinary Medical Association. 1994;205:968.

[24] Bermejo M, Rodríguez-Teijeiro JD, Illera G, Barroso A, Vilà C, Walsh PD. Ebola outbreak killed 5000 gorillas. Science. 2006;314:1522–1523. DOI: 10.1126/science.1133105

[25] De Carli G, Fusco FM, Puro V, Ippolito G. Adopting a global safety standard for the prevention of Ebola needle-stick exposures. Infection Control and Hospital Epidemiology. 2015;36:745–746. DOI: 10.1017/ice.2015.62.

[26] European Biosafety Network, 2013. European Healthcare Workers say major problems still need to be addressed to prevent sharps injuries. http://www.europeanbiosafety-network.eu/news#item-127. Accessed 15.08.16.

[27] Wilkins JL, Bowman M. Needle stick injuries among female veterinarians: frequency, syringe contents, and side effects. Occupational Medicine. 1997;47:451–457.

[28] Patterson C, LaVenture M, Hurley S, Davis J. Accidental self-inoculation with *Mycobacterium paratuberculosis* bacterin (Johne's bacterin) by veterinarians in Wisconsin. Journal of the American Veterinary Medical Association. 1988;192:1197–1199.

[29] Hill DJ, Langley RL, Morrow WM. Occupational injuries and illnesses reported by zoo veterinarians in the United States. Journal of Zoo Animal Medicine. 1998;29:371–385.

[30] Leggat PA, Smith DR, Speare R. Exposure rate of needlestick and sharps injuries among Australian veterinarians. Journal of Occupational Medicine and Toxicology. 2009;28:25. DOI: 10.1186/1745-6673-4-25.

[31] Whitney EA, Ailes E, Myers LM, Saliki JT, Berkelman RL. Prevalence of and risk factors for serum antibodies against Leptospira serovars in US veterinarians. Journal of the American Veterinary Medical Association. 2009;234(7):938–944. DOI: 10.2460/javma.234.7.938.

[32] Kabuusu RM, Keku EO, Kiyini R, McCann TJ. Prevalence and patterns of self-reported animal-related injury among veterinarians in metropolitan Kampala. Journal of Veterinary Science. 2010;11(4):363–365.

[33] Mesquita JR, Sousa SI, Vala H, Nascimento MS. The epidemiology of blood-contaminated needlestick injuries among veterinarians in Portugal. Journal of Agromedicine. 2015;20(2):160–166. DOI: 10.1080/1059924X.2015.1010061

[34] Muroga N, Yamamoto T, Hayama Y, Kobayashi S, Hidano A, Tsutsui T. Injuries to staff engaged in foot-and-mouth disease eradication in Japan. Occupational Medicine (Oxford, England). 2015;65(1):45–48. DOI: 10.1093/occmed/kqu179.

[35] Seng M, Lim JW, Sng J, Kong WY, Koh D. Incidence of needlestick injuries among medical students after implementation of preventive training. Singapore Medical Journal. 2013;54(9):496–500.

[36] Bhardwaj A, Sivapathasundaram N, Yusof M, Minghat A, Swe K, Sinha N. The prevalence of accidental needle stick injury and their reporting among healthcare workers in orthopaedic wards in general hospital Melaka, Malaysia. Malaysian Orthopaedic Journal. 2014;8(2):6–13. DOI: 10.5704/MOJ.1407.009.

[37] European Biosafety Network, 2011. Toolkit for the implementation of European Directive on Prevention from sharps injuries. http://www.europeanbiosafetynetwork.eu/EBN%20Toolkit%202011.pdf. Accessed 16.08.16.

[38] Ream PS, Tipple AF, Barros DX, Souza AC, Pereira MS.Biological risk among hospital housekeepers. Archives of Environmental and Occupational Health. 2016;71(2):59–65. DOI: 10.1080/19338244.2014.927347.

[39] De Carli G, Abiteboul D, Puro V. The importance of implementing safe sharps practices in the laboratory setting in Europe. Biochemia Medica. 2014;24:45–56. DOI: 10.11613/BM.2014.007.

[40] Jayanth ST, Kirupakaran H, Brahmadathan KN, Gnanaraj L, Kang G. Needle stick injuries in a tertiary care hospital. Indian Journal of Medical Microbiology. 2009;27(1): 44–47.

[41] bin Zakaria M, Lerche N, Chomel B, Kass P. Accidental injuries associated with nonhuman primate exposure at two regional primate research centers (USA): 1988–1993. Laboratory Animal Science. 1996;46:298–304.

[42] EpiNet. International Healthcare Worker Safety Center, University of Virginia: Summary report for needlestick and sharp object injuries reporting period: Jan 01, 2000 to Dec 31; 2006. Accessed 21.08.16.

[43] Cho E, Lee H, Choi M, Park SH, Yoo IY, Aiken LH. Factors associated with needlestick and sharps injuries among hospital nurses: a cross-sectional questionnaire survey. International Journal of Nursing Studies. 2013;50(8):1025–1032. DOI: 10.1016/j.ijnurstu.2012.07.009.

[44] Martins A, Coelho AC, Vieira M, Matos M, Pinto ML. Age and years in practice as factors associated with needlestick and sharps injuries among health care workers in a Portuguese hospital. Accident Analysis & Prevention. 2012;47:11–15. DOI: 10.1016/j.aap.2012.01.011.

[45] Adib-Hajbaghery M, Lotfi MS. Behavior of healthcare workers after injuries from sharp instruments. Trauma Monthly. 2013;18:75–80. DOI:10.5812/traumamon.12779.

[46] Epp T, Waldner C. Occupational health hazards in veterinary medicine: physical, psychological, and chemical hazards. The Canadian Veterinary Journal. 2012;53:151–157.

[47] Nienhaus A, Skudlik C, Seidler A. Work-related accidents and occupational diseases in veterinarians and their staff. International Archives of Occupational and Environmental Health. 2005;78(3):230–238. DOI: 10.1007/s00420-004-0583-5.

[48] Landercasper J, Cogbill TH, Strutt PJ, Landercasper BO. Trauma and the veterinarian. Journal of Trauma. 1988;28:1255–1259.

[49] Gabel CL, Gerberich SG. Risk factors for injury among veterinarians. Epidemiology. 2002;13:80–86.

[50] Lindahl C, Lundqvist P, Hagevoort GR, Lunner Kolstrup C, Douphrate DI, Pinzke S, Grandin T. Occupational health and safety aspects of animal handling in dairy production. Journal of Agromedicine. 2013;18(3):274–283. DOI: 10.1080/1059924X.2013.796906.

[51] Watson K, Williams-Johnson J, Watson H, Walters C, Williams EW, Eldemire-Shearer D. The compliance of healthcare workers with universal precautions in the emergency room at the university hospital of the west indies. West Indian Medical Journal. 2014;63(3):217–225. DOI: 10.7727/wimj.2014.044.

[52] Hafer AL, Langley RL, Morrow WM, Tulis JJ. Occupational hazards reported by swine veterinarians in the United States. Swine Health and Production. 1996;4:128–141.

[53] D'Souza E, Barraclough R, Fishwick D, Curran A. Management of occupational health risks in small-animal veterinary practices. Occupational Medicine. 2009;59:316–322. DOI: 10.1093/occmed/kqn125.

[54] Scheftel JM, Elchos BL, Cherry B, DeBess EE, Hopkins SG, Levine JF, Williams CJ, Bell MR, Dvorak GD, Funk RH, Just SD, Samples OM, Schaefer EC, Silvia CA. Compendium of veterinary standard precautions for zoonotic disease prevention in veterinary personnel: National Association of State Public Health Veterinarians Veterinary Infection Control Committee 2010. Journal of the American Veterinary Medical Association. 2010;237:1403–1422. DOI: 10.2460/javma.237.12.1403.

[55] Australian Veterinary Association. Guidelines for Veterinary Personal Biosecurity. 2. St. Leonards, NSW: Australian Veterinary Association; 2013.

[56] Tavasoly A, Moosavi Z, Bazargani T, Bazargani M. Accidental self-inoculation in a veterinarian with attenuated vaccine of bovine Johne's disease. Iranian Journal of Veterinary Research. 2007;8(4):374–376.

[57] Thompson RN, McNicholl BP. Needlestick and infection with horse vaccine. BMJ Case Reports. 2010; bcr1120092444.

[58] Ashford DA, di Pietra J, Lingappa J, Woods C, Noll H, Neville B, Weyant R, Bragg SL, Spiegel RA, Tappero J, Perkins BA. Adverse events in humans associated with accidental exposure to the livestock brucellosis vaccine RB51. Vaccine. 2014;22:3435–3439. DOI: 10.1016/j.vaccine.2004.02.041.

[59] Wilkins MJ, Bartlett PC, Judge LJ, Erskine RJ, Boulton ML, Kaneene JB. Veterinarian injuries associated with bovine TB testing livestock in Michigan, 2001. Preventive Veterinary Medicine. 2009;89(3–4):185–190. DOI: 10.1016/j.prevetmed.2009.02.014

[60] Wicker S, Stirn AV, Rabenau HF, von Gierke L, Wutzler S, Stephan C. Needlestick injuries: causes, preventability and psychological impact. Infection. 2014;42(3):549–552. DOI: 10.1007/s15010-014-0598-0.

[61] Wald J. The psychological consequences of occupational blood and body fluid exposure injuries. Disability and Rehabilitation. 2009;31:1963–1969.

[62] Green B, Griffiths EC. Psychiatric consequences of needlestick injury. Occupational Medicine. 2013;63:183–188. DOI: 10.1093/occmed/kqt006

[63] Lucas M, Day L, Shirangi A, Fritschi L. Significant injuries in Australian veterinarians and use of safety precautions. Occupational Medicine. 2009;59(5):327–333. DOI: 10.1093/occmed/kqp070.

[64] Jones DP. Accidental self inoculation with oil based veterinary vaccines. The New Zealand Medical Journal. 1996;109(1030):363–365.

[65] Jahan S. Epidemiology of needlestick injuries among health care workers in a secondary care hospital in Saudi Arabia. Annals of Saudi Medicine. 2005;25(3):233–238.

[66] Laboratory Centre for Disease Control. Preventing the transmission of bloodborne pathogens in health care and public service settings. Canada Communicable Disease Report. 1997;23(Suppl 3):i–vii,1–43, i–vii, 1–52.

[67] Trim JC, Elliott TS. A review of sharps injuries and preventative strategies. Journal of Hospital Infection. 2003;53:237–242.

[68] Panlilio AL, Orelien JG, Srivastava PU, Jagger J, Cohn RD, Cardo DM. Estimate of the annual number of percutaneous injuries among hospital-based health care workers in the United States, 1997–1998. Infection Control and Hospital Epidemiology. 2004;25:556–562. DOI: 10.1086/502439.

[69] Centers for Disease Control and Prevention. Guidelines for prevention of human immunodeficiency virus and hepatitis B virus to health-care and public-safety workers. Morbidity and Mortality Weekly Report. 1989;38(Suppl 6):111–155.

[70] Thomas S, Agarwal M, Mehta G. Intraoperative glove perforation—single versus double gloving in protection against skin contamination. Postgraduate Medical Journal. 2001;77(909):458–460.

[71] Lancaster C, Duff P. Single versus double-gloving for obstetric and gynecologic procedures. American Journal of Obstetrics and Gynecology. 2007;196(5):e36–e37.

[72] Yang L, Mullan B. Reducing needle stick injuries in healthcare occupations: an integrative review of the literature. ISRN Nursing. 2011;2011:315432. DOI: 10.5402/2011/315432.

[73] Lavoie MC, Verbeek JH, Pahwa M. Devices for preventing percutaneous exposure injuries caused by needles in healthcare personnel. Cochrane Database the Cochrane Database of Systematic Reviews. 2014;(3):CD009740. DOI: 10.1002/14651858.CD009740.pub2.

[74] Council of the European Union. Council Directive 2010/32/EU of 10 May 2010 implementing the Framework Agreement on prevention from sharps injuries in the hospital and healthcare sector concluded by HOSPEEM and EPSU. Official Journal of the European Union. 2010;L134(53):66–72. Available at: https://osha.europa.eu/pt/legislation/directives/council-directive-2010-32-eu-prevention-from-sharp-injuries-in-the-hospital-and-healthcare-sector. Accessed 27.08.16.

[75] Loudon MA, Stonebridge PA. Minimizing the risk of penetrating injury to surgical staff in the operating theatre: towards sharp-free surgery. Journal of the Royal College of Surgeons of Edinburgh. 1998;43:6–8.

[76] United States General Accounting Office. Occupational Safety: Selected Cost and Benefit Implications of Needlestick Prevention Devices for Hospitals. GAO-01-60R; November 17, 2000.

[77] Smith R, Stilts P. Controlling workers' compensation losses. Journal of the American Veterinary Medical Association. 1996;209:526.

[78] Pruss-Ustun A, Rapiti E, Hutin Y. Estimation of the global burden of disease attributable to contaminated sharps injuries among health-care workers. American Journal of Industrial Medicine. 2005;48(6):482–490. DOI: 10.1002/ajim.20230.

[79] Hanmore E, Maclaine G, Garin F, Alonso A, Leroy N, Ruff L. Economic benefits of safety-engineered sharp devices in Belgium–a budget impact model. BMC Health Services Research. 2013;13:489. DOI: 10.1186/1472-6963-13-489.

[80] Siegel JD, Rhinehart E, Jackson M, Chiarello L. 2007 Guideline for isolation precautions: preventing transmission of infectious agents in health care settings. American Journal of Infection Control. 2007;35(10 Suppl 2):S65–S164. DOI: 10.1016/j.ajic.2007.10.007.

[81] Riley CB, McCallum S, MacDonald JA, Hill KE. A prospective observational study of needle-handling practices at a University Veterinary Teaching Hospital. New Zealand Veterinary Journal. 2016;64:117–120. DOI: 10.1080/00480169.2015.1100100.

Chapter 8

Gene-Environment Interactions: The Case of Asbestosis

Vita Dolzan, Metoda Dodic-Fikfak and
Alenka Franko

Additional information is available at the end of the chapter

http://dx.doi.org/10.5772/66120

Abstract

It is becoming evident that both environmental/lifestyle and genetic factors may influence the development of many diseases. This chapter highlights the importance of considering gene-environment interactions, which is shown on the example of our studies into asbestosis, one of the most frequent asbestos-related diseases. Asbestos fibres induce generation of reactive oxygen and nitric species (ROS and RNS), and it is generally accepted that ROS and RNS are involved in the pathogenesis of asbestos-related diseases. Human tissues contain specific enzymes that metabolise ROS and RNS, such as superoxide dismutases (SODs), catalase (CAT), glutathione-S-transferases (GSTs) and inducible nitric oxide synthase (iNOS). As these enzymes are encoded by polymorphic genes, genetic variability in an individual's capacity to detoxify these reactive species may modify the risk for disease. Our previous studies into asbestosis showed that the associations between the risk of asbestosis and *MnSOD* Ala-9Val polymorphism and between asbestosis and *iNOS* genotypes were modified by *CAT* −262C>T polymorphism. A strong interaction was also found between smoking (lifestyle factor) and *GSTM1*-null polymorphism, between smoking and *iNOS* $(CCTTT)_n$ polymorphism and between cumulative asbestos exposure (environmental factor) and *iNOS* $(CCTTT)_n$ polymorphism. The findings of our studies and other studies indicate that in addition to environmental and/or occupational exposure to different hazards and lifestyle factors, genetic factors as well as the interactions between different genotypes, between genotypes and lifestyle factors and between genotypes and environmental/occupational exposure to hazards may also have an important role on the development of diseases and should be further investigated.

Keywords: asbestosis, exposure, gene-environment interactions, gene-gene interactions

1. Introduction

It is becoming evident that both environmental and genetic factors may influence the development of many diseases [1–7]. It is therefore important to consider gene-environment interactions when studying diseases related to exposure to different hazards and lifestyle factors.

Environmental and lifestyle factors have been investigated in many epidemiological studies using self-reported information obtained by questionnaires, interviews, records or measurements of exposure. However, very few epidemiological studies included the information on genetic risk factors. Similarly, many studies investigating genetic factors obtained little information on environmental factors and lifestyle. Genetic predisposition can be presumed from family history, from phenotypic characteristics (e.g. metabolic capacity) or, most importantly, from an analysis of deoxyribonucleic acid (DNA) sequence [8].

The research into gene-environment interactions requires the information on both environmental/lifestyle and genetic factors [7, 8]. Primary candidates for gene-environment interaction studies have been mostly genes coding for xenobiotic-metabolising enzymes [3]. Genetic variability in these genes may lead to interindividual differences in the capacity for xenobiotic metabolism, thus modifying an individual's susceptibility to the development of disease [3].

The approach to the analysis of gene-environment interactions is presented using the example of our study into asbestosis, which is one of the most frequent asbestos-related diseases. According to the model of causation, asbestos exposure, genetic factors and possibly also unknown causes have a crucial role in the occurrence of asbestosis [9]. Although asbestos-related diseases are among the most extensively studied occupational diseases, and the causal relationship between asbestos exposure and asbestosis has been well proved [10–14], relatively little has been known about the genetic factors that might modify an individual's susceptibility to the development of this disease [6, 15–17].

2. Asbestos exposure

Asbestos is a commercial name for a group of fibrous silicates with certain toxic properties, such as the ability to produce inflammation, fibrous scarring and cancer [18–20]. Based on their physical and chemical structures, asbestos fibres can be classified into two major groups: chrysotile and amphiboles [20–25].

Occupational exposure to asbestos occurs in asbestos mining, production and milling of asbestos fibres; in asbestos cement industry; in construction; in machine and insulation product industry; in ship building or repair; in car industry; in production of brakes and clutches; in car, bus, lorry, railway carriage and aeroplane repair; in asphalt mixing; in disposal of asbestos waste and materials; in brickworks; in textile industry and in other industries and activities [20, 22, 26–28].

Local population can be exposed to asbestos mostly in the neighbourhood of factories where asbestos is produced or used (exposure to polluted air, water and food). The source of environmental asbestos exposure may also be asbestos cement sheets, asbestos insulators and other asbestos-containing products. Asbestos fibres may also be found in water which flows through asbestos cement pipes, especially if they have been damaged. Workers exposed to asbestos may bring asbestos home to the family members on clothing or hair [26–28].

3. Asbestos-related diseases

Asbestos exposure has been associated with the development of asbestosis; pleural diseases, such as pleural plaques, diffuse pleural thickening and pleural effusion and several types of cancer: lung cancer, diffuse malignant mesothelioma of the pleura and peritoneum, cancer of the larynx, cancer of the ovary as well as the cancers of the buccal mucosa, the pharynx, the gastrointestinal tract and the kidney [11, 12, 16, 25, 29–41].

4. Clinical presentation of asbestosis

Asbestosis is an interstitial pulmonary process that develops into diffuse pulmonary fibrosis after a long latency period [42, 43]. The disease continues to progress even after the cessation of exposure, and the process is irreversible. One of the earliest symptoms may be dyspnoea, which is manifested at first only after strenuous exertion, but subsequently with less and less exertion, and eventually it appears even at rest. Another non-specific symptom and usually late manifestation of the disease is irritating and dry, usually non-productive cough, sometimes associated with chest pain [42, 44]. Pulmonary function changes are characterised mostly by a restrictive impairment [27, 28, 42–44]. Later, obstructive airway impairment may also occur [27, 28]. On chest radiographs, small irregular opacities appear initially in the lower lung fields that may enlarge with more advanced disease and involve also middle lung fields [27, 42–44]. Characteristic features of asbestosis on high-resolution computed tomography (HRCT) include fibrotic intralobular interstitial thickening and interlobular septal thickening, subpleural lines and opacities, parenchymal bands, ground-glass opacities and, in more severe disease, variable honeycombing [27].

5. Reactive oxygen and nitric species: the link between asbestos exposure and the development of asbestosis

The pathogenesis of asbestosis is still poorly understood. The findings of studies on cell cultures and animal models indicate that reactive oxygen and nitric species (ROS and RNS) are involved in the pathogenesis of this disease [23, 30, 45–55]. The most important reactive metabolites in the pathogenesis of asbestos-related lung diseases are superoxide anion (O_2^-), hydrogen peroxide (H_2O_2), hydroxyl radical ($OH^{\cdot}$) and nitric oxide (NO) [46, 48, 56, 57]. Asbestos may stimulate the production of ROS in two different ways. The first mechanism involves redox-active iron (Fe^{2+}, Fe^{3+}) in asbestos that catalyses the formation of $OH^{\cdot}$, whereas the second mechanism involves the production of ROS by alveolar macrophages during the phagocytosis of asbestos fibres [58–60]. Reactive oxygen species in lungs may lead to the production of cytotoxic and potentially genotoxic electrophilic compounds [46].

It has also been suggested that asbestos fibres may upregulate the activity of inducible nitric oxide synthase (iNOS), thus inducing the production of NO by alveolar macrophages and pulmonary epithelial cells [51, 61–64]. Because NO is a free radical, it reacts readily with other reactive oxygen metabolites (as, for instance, O_2^-), leading to the formation of toxic metabolites, most importantly peroxynitrite [65–69]. Nitric oxide may play a role in the initiation and progression of asbestosis [51, 64, 70, 71]. However, the data presented by Dörger et al. [72] indicate that iNOS-derived NO plays a dual role in acute asbestos-induced lung injury and that although iNOS deficiency resulted in an exacerbated inflammatory response, it improved oxidant-promoted lung tissue damage.

Reactive oxygen species and RNS can damage all types of biomolecule, including lipids, proteins and deoxyribonucleic acid (DNA). Complex defence mechanisms, including enzymes, proteins and antioxidants, are involved in the prevention of cell damage [73, 74].

6. Enzymes involved in the detoxification of reactive oxygen and nitric species

Human tissues contain specific enzyme systems to detoxify ROS and RNS. Superoxide dismutases (SODs) and catalase (CAT) together with glutathione peroxidases represent an important line of the primary antioxidant enzyme defence system against ROS. Superoxide dismutases catalyse the dismutation of O_2^- to H_2O_2 and oxygen (O_2), whereas CAT subsequently catalyses the conversion of H_2O_2 to water (H_2O) and O_2 [48, 75–82]. Three distinct SOD isoenzymes have been identified in mammals: a cytosolic copper-zinc SOD (CuZnSOD or SOD1) localised in cytoplasmic compartment with cooper (Cu) and zinc (Zn) in the catalytic centre, manganese SOD (MnSOD or SOD2) that is localised in mitochondria and uses manganese (Mn) as a cofactor and extracellular SOD (ECSOD or SOD3) that also contains Cu and Zn in the catalytic centre and is located in the extracellular space [74, 82, 83].

Another important family of enzymes involved in the detoxification of xenobiotics and electrophiles produced by ROS and RNS is glutathione S-transferases (GSTs) [84–87]. They catalyse the conjugation of reduced glutathione to different electrophiles [88]. These conjugation reactions mostly result in less reactive products [89]; however, in some cases, the products are more reactive and consequently more harmful than the parent compound [90, 91]. Seven classes of cytosolic GST isoenzymes have been recognised in mammals (Alpha, Mu, Pi, Sigma, Theta, Omega, Zeta) [84–86, 91, 92]. The major GST enzyme in the human lung is GSTP1, which belongs to the Pi class [90, 91, 93], while GSTM1 (Mu class) and GSTT1 (Theta class) were most frequently investigated [90, 91].

7. Genetic variability of metabolic enzymes

Genetic polymorphisms are the most common cause for genetic variability of detoxification and antioxidative enzymes [15–17, 80, 91, 94–99].

The most common functional single nucleotide polymorphism (SNP) of the *MnSOD* gene is C to T substitution (c.201C>T, rs4880), which results in alanine (Ala) to valine (Val) amino acid change at position −9 of the mitochondrial targeting sequence (MnSOD p.Ala-9Val) [96, 97, 100]. It has been suggested that this SNP alters the secondary structure of the protein and hence may affect the efficiency of transport of the MnSOD into the mitochondria, where it would be biologically available [96, 97].

ECSOD is secreted into extracellular space where it binds lung matrix components and inhibits their fragmentation in response to oxidative stress [101, 102]. In the *ECSOD* gene, a C to G substitution (c.896C>G, rs1799895) leads to amino acid change from arginine (Arg) to glycine (Gly) at position 213 (p.Arg213Gly) [89, 100, 103–105]. This polymorphism causes an 8- to 15-fold increase in the concentration of plasma ECSOD levels due to impaired binding to the extracellular matrix [103, 104].

The most common functional SNP of the catalase gene (*CAT*) consists of a C to T substitution at position −262 in the promoter region (*CAT* c.−262C>T) and has a substantial impact on the basal expression as well as the CAT levels in red blood cell [80]. The findings of later studies indicated lower CAT activity in subjects with the −262TT genotype than those with the CT and CC genotypes [106–111].

Regarding GSTs, the most common polymorphism of the *GSTM1* and *GSTT1* genes in most of the populations is null polymorphism due to homozygous deletion (null genotype) of these genes, which result in the absence of the GSTM1 and GSTT1 enzyme activity [17, 91]. *GSTM1*-null genotype has been associated with an increased risk of asbestosis in some studies [16, 86], while this association has not been proved in the others [15, 17]. No association has been found between *GSTT1* deletion polymorphism and asbestosis in the studies published so far [17, 86]. As for the *GSTP1* gene, two common single nucleotide polymorphisms in the coding sequence were reported to result in amino acid substitution that may lead to reduced conjugating activity of the enzyme [91, 98, 112, 113]. The first polymorphism is characterised by adenine (A) to guanine (G) transition of nucleotide 313 in exon 5 (c.A313G), which causes an isoleucine (Ile) to valine (Val) substitution at position 105 of the GSTP1 enzyme (p.Ile105Val), resulting in three possible genotypes: 105 Ile/Ile, 105 Ile/Val or 105 Val/Val. The second polymorphism involves the cytosine (C) to thymine (T) transition at nucleotide 341 in exon 6 (c.C341T), which results in alanine (Ala) to Val substitution at position 114 of the GSTP1 enzyme (p.Ala114Val). Regarding codon 114, three genotypes are also possible: 114 Ala/Ala, 114 Ala/Val or 114 Val/Val [91, 98]. Based on the presence of the polymorphisms in both codons 105 and 114, *GSTP1* genotypes can be combined into groups with a presumed high, intermediate or low conjugation capacity of the enzyme.

The human *iNOS* gene is also known to be polymorphic. Several types of polymorphisms have been identified in the promoter region of the *iNOS* gene [99, 114]. The CCTTT pentanucleotide tandem repeat polymorphisms have been associated with the transcriptional promoter activity, which has been shown to increase with the CCTTT repeat number. Based on that, alleles with 11 or fewer CCTTT repeats are usually defined as short alleles (S) and the ones with 12 or more repeats as long alleles (L). Accordingly, the subjects can have SS, SL or LL genotype [115].

8. Gene-environment interactions and asbestosis

We are presenting the example of an approach to gene-environment interaction research by summarising and building on the results of our studies that aimed to investigate the influence of interactions between different genotypes (*MnSOD, ECSOD, CAT, GSTM1, GSTT1, GSTP, iNOS*), between genotypes and smoking and between genotypes and cumulative asbestos exposure on the risk of developing asbestosis [6, 14, 116–119].

A nested case-control study included 262 cases with asbestosis and 265 controls with no asbestos-related disease. All the subjects included in the study were employed in the asbestos cement manufacturing plant of Salonit Anhovo, Slovenia, and occupationally exposed to asbestos. Data on smoking were obtained from all subjects using a standardised questionnaire [25, 120] and checked during the interview. The data on the cumulative asbestos exposure, expressed in fibres/cm^3-years [intensity in fibres per cm^3 of air multiplied by time of exposure expressed in years], were available for all the subjects from the previous study [25]. The diagnosis of asbestosis or 'no asbestos-related disease' was based on the Helsinki Criteria for Diagnosis and Attribution of Asbestos Diseases [121] and on the American Thoracic Society recommendations [122]. Each case was confirmed by an interdisciplinary group of experts (consisting of an occupational physician, a radiologist and a pulmonologist) of the State Board for Recognition of Occupational Asbestos Diseases at the Clinical Institute of Occupational Medicine. Capillary blood samples from the finger tips of all cases and controls have been collected on FTA Mini Cards (Whatman Bioscience) for the isolation of deoxyribonucleic acid (DNA) and genotyping. All the genetic analyses were performed using PCR-based approaches as previously described [6, 14, 116–119].

Before testing interactions, the associations between outcome [in our case asbestosis] and individual variables were assessed using univariate logistic regression analysis. As expected, asbestosis was associated with cumulative asbestos exposure, whereas no association was found with smoking (OR = 0.98, 95%; CI = 0.69–1.39 for ever versus never smoking) [14]. Analysing the association between asbestosis and individual genotypes, an important association was observed between asbestosis and *MnSOD* genotype (OR = 1.50, 95% CI = 1.01–2.24 for -9Ala/Ala versus combined Ala/Val and Val/Val genotypes) [118]. Only non-significantly elevated risk of asbestosis was observed for the *ECSOD* and *CAT* genotypes (OR = 1.63, 95% CI = 0.62–4.27 for *ECSOD* 213Arg/Gly versus the Arg/Arg genotype and OR = 1.36, 95% CI = 0.70–2.62 for *CAT* −262 TT compared to combined CT and CC genotypes, respectively) [117, 118]. Regarding GSTs, no association was found between asbestosis and *GSTM1*-null genotype (OR = 1.01, 95% CI = 0.71–1.43), while the presence of *GSTT1*-null genotype showed a protective effect for this disease (OR = 0.61, 95% CI = 0.40–0.94) [14]. On the other hand, *GSTP1* genotype coding for an enzyme with a high conjugation capacity versus genotypes resulting in an intermediate or low enzyme activity significantly increases the risk of developing asbestosis (OR = 1.49, 95% CI 1.06–2.10) [116]. A slightly elevated risk of asbestosis was also found for the *iNOS* LL genotype compared to the combined SL and SS genotypes (OR = 1.20, 95% CI = 0.85–1.69) [119]. Based on the above-mentioned results, it could be suggested that the genotypes may increase, decrease or have no effect on the risk of disease, in our case asbestosis.

Univariate modelling was followed by multivariate analysis and interactions as the genes usually do not act independently, but may interact. To test the interactions, simple categorical models based on stratification were constructed first, followed by logistic regression models using dummy variables. The analysis showed that the association between asbestosis and *MnSOD* Ala-9Val genotypes was modified strongly by *CAT* −262 C>T genotypes. An increased risk of developing asbestosis was observed for the combined *MnSOD* -9Ala/Val and Val/Val genotypes compared to the Ala/Ala genotype only among those subjects who also had *CAT* −262TT genotype, suggesting an interaction, which was further confirmed by logistic regression analysis using dummy variables (OR = 4.49, 95% CI = 1.08–18.61) [6]. Considering that both MnSOD and CAT constitute a part of the primary defence system against ROS and catalyse the consecutive reactions in the detoxification of ROS [48, 74, 80, 82], this interaction could be considered as logical and biologically plausible. Similarly, the association between asbestosis and *iNOS* $(CCTTT)_n$ genotypes was also modified by CAT −262 C>T genotypes, where a higher asbestosis risk for the *iNOS* LL genotype versus the combined SL and SS genotypes was observed only among those who had *CAT* −262 TT genotype (OR = 4.78, 95% CI = 1.15–19.81) [6]. Taking into account that reactions between ROS and NO have been proposed to potentiate the cytotoxic and mutagenic effect of asbestos fibres [48, 51, 64, 71] and based on the assumption that NO produced by the catalytic activity of iNOS can function as a protective agent against toxic effects of H_2O_2 [123], which is detoxified by CAT [48, 74, 80, 82], and vice versa that H_2O_2 decreases the cytotoxicity of NO [124], this interaction could also be considered as biologically plausible [6].

Next, interactions between different genotypes and an important lifestyle factor—in our case smoking—have been tested. We observed that the *GSTM1*-null polymorphism did modify the association between smoking and asbestosis, although there was no independent association between either *GSTM1*-null polymorphism or smoking and asbestosis risk (OR = 2.67, 95% CI = 1.31–5.46) [6]. We can explain this modifying effect with the observation that both asbestos and smoking increase the production of ROS [46, 125, 126], which are known to be involved in the pathogenesis of asbestosis [23, 30, 46, 48–50]. It has been suggested that cigarette smoke and asbestos increase DNA damage and ROS production in pulmonary cells synergistically [125–127]. In line with these reports and considering the role of *GSTM1* in the defence against ROS [84–87], this observation could also be considered as biologically plausible [6]. Similarly, the association between smoking and asbestosis was modified by *iNOS* $(CCTTT)_n$ polymorphism (OR = 2.00, 95% CI = 0.99–4.03) [6]. Knowing that cigarette smoke is the largest source of NO that humans are exposed to and can also increase the expression and activity of iNOS [128, 129] and based on the suggestion that asbestos fibres may upregulate the activity of iNOS and thus the production of NO, which is thought to play an important role in the initiation and progression of asbestosis [51, 70], this interaction could also be physiologically explained [6].

Finally, we present an example of the interaction between genotypes and environmental exposure, in our case occupational exposure to asbestos. In order to assess the interactions between the genotypes and occupational cumulative asbestos exposure, we have first constructed simple categorical models that included cumulative asbestos exposure categorised as follows: ≤11.23 fibres/cm^3-years and >11.23 fibres/cm^3-years (11.23 fibres/cm^3-years was the average cumulative asbestos exposure for the controls). In our analysis, we have observed that the association between asbestosis and cumulative asbestos exposure was modified by the iNOS $(CCTTT)_n$ genotypes (OR = 5.74; 95% CI = 3.30–9.99) [6].

9. Conclusions

The findings of our studies suggest that in addition to environmental and/or occupational exposure to different hazards and lifestyle factors, the genetic factors and the interactions between different genotypes, between genotypes and lifestyle factors and between genotypes and environmental/occupational exposure to hazards may have an important influence on the development of diseases and should be further investigated [6, 130–133]. In agreement with our observations, an increasing number of molecular epidemiological studies support the importance of investigating not only genetic predisposition but also gene-gene and gene-environment interactions when assessing the risk of developing diseases [134–136]. Novel high-throughput technologies may also allow the investigation of interactions between exposure to hazards and epigenetic changes in disease risk assessment [137].

Acknowledgement

This work was financially supported by the Slovenian Research Agency [grants L3-3648 and P1-0170].

Author details

Vita Dolzan[1], Metoda Dodic-Fikfak[2] and Alenka Franko[2*]

*Address all correspondence to: alenka.franko@siol.net

1 Pharmacogenetics Laboratory, Institute of Biochemistry, Faculty of Medicine, University of Ljubljana, Slovenia

2 Clinical Institute of Occupational Medicine, University Medical Centre, Ljubljana, Slovenia

References

[1] Greenland S, Rothman KJ. Concepts of interaction. In: Rothman KJ, Greenland S, editors. Modern epidemiology Philadelphia, PA: Lippincott-Raven, 1998. pp. 329–342.

[2] Furberg AH, Ambrosone CB. Molecular epidemiology, biomarkers and cancer prevention. Trends Mol Med. 2001;7:517–521. PubMed PMID: 11689338.

[3] Mucci LA, Wedren S, Tamimi RM, Trichopoulos D, Adami HO. The role of gene-environment interaction in the aetiology of human cancer: examples from cancers of the large bowel, lung and breast. J Intern Med. 2001;249:477–493. PubMed PMID: 11422654.

[4] Boks MP, Schipper M, Schubart CD, Sommer IE, Kahn RS, Ophoff RA. Investigating gene environment interaction in complex diseases: increasing power by selective sampling for environmental exposure. Int J Epidemiol. 2007;36:1363–1369. PubMed PMID: 17971387.

[5] Khoury MJ, Millikan R, Gwinn M. Genetic and molecular epidemiology. In: Rothman KJ, Greenland S, Lash TL editors. Modern Epidemiology Philadelphia, Baltimore, New York, London, Buenos Aires, Hong Kong, Sydney, Tokyo: Lippincott-Raven, 2008. pp. 564–597.

[6] Franko A, Dolžan V, Arnerić N, Dodič-Fikfak M. The influence of gene-gene and gene-environment interactions on the risk of asbestosis. Biomed Res Int. 2013;2013:405743. PubMed PMID: 23984360; doi: 10.1155/2013/405743.

[7] Lu M, Lee HS, Hadley D, Huang JZ, Qian X. Logistic principal component analysis for rare variants in gene-environment interaction analysis. IEEE/ACM Trans Comput Biol Bioinform. 2014;1:1020–1028. PubMed PMID: 26357039; doi: 10.1109/TCBB.2014.2322371.

[8] Hunter DJ. Gene-environment interactions in human diseases. Nat Rev Genet. 2005;6:287–298. PubMed PMID: 15803198.

[9] Rothman KJ, Greenland S, Poole C, Lash TL. Causation and cause inference. In: Rothman KJ, Greenland S, Lash TL editors. Modern Epidemiology Philadelphia, Baltimore, New York, London, Buenos Aires, Hong Kong, Sydney, Tokyo: Lippincott-Raven, 2008. pp. 5–22.

[10] Mossman BT, Gee JB. Asbestos-related diseases. N Engl J Med. 1989;320:1721–1730. PubMed PMID: 2659987.

[11] Dement JM, Brown DP, Okun A. Follow-up study of chrysotile asbestos textile workers: cohort mortality and case–control analyses. Am J Ind Med. 1994;26:431–447. PubMed PMID: 7810543.

[12] Dement JM, Brown DP. Lung cancer mortality among asbestos textile workers: a review and update. Ann Occup Hyg. 1994;38:525–532. PubMed PMID: 7978974.

[13] Jakobsson K, Strömberg U, Albin M, Welinder H, Hagmar L. Radiological changes in asbestos cement workers. Occup Environ Med. 1995;52:20–27. PubMed PMID: 7697136.

[14] Franko A, Dodič-Fikfak M, Arnerić N, Dolžan V. Glutathione S- transferases GSTM1 and GSTT1 polymorphisms and asbestosis. J Occup Environ Med. 2007;49:667–671. PubMed PMID: 17563610.

[15] Jakobsson K, Rannug A, Alexandrie AK, Rylander L, Albin M, Hagmar L. Genetic polymorphism for glutathione-S-transferase mu in asbestos cement workers. Occup Environ Med. 1994;51:812–816. PubMed PMID: 7849864.

[16] Smith CM, Kelsey KT, Wiencke JK, Leyden K, Levin S, Christiani DC. Inherited glutathione-S-transferase deficiency is a risk factor for pulmonary asbestosis. Cancer Epidemiol Biomarkers Prev. 1994;3:471–477. PubMed PMID: 8000297.

[17] Hirvonen A, Saarikoski ST, Linnainmaa K, Koskinen K, Husgafvel-Pursiainen H, Mattson K, Vainio H. Glutathione S-transferase and N-acetyltransferase genotypes and asbestos-associated pulmonary disorders. J Natl Cancer Inst. 1996;88:1853–6185. PubMed PMID: 8961976.

[18] Becklake MR. Exposure to asbestos and human disease. N Engl J Med. 1982; 06:1480–1482. PubMed PMID: 7078593.

[19] Craighead JE, Mossman BT. The pathogenesis of asbestos-associated diseases. N Engl J Med. 1982;306:1446–1455. PubMed PMID: 7043267.

[20] Wagner GR, Hearl FJ. Mineral dust: asbestos, silica, coal, manufactured fibers. In: Rosenstock L, Cullen MR, Brodkin CA, Redlich CA, editors. Textbook of clinical occupational and environmental medicine. 2nd ed. Philadelphia, Edinburgh, New York, St Louis, Sydney, Toronto: Elsevier Saunders, 2005. pp. 1073–1078.

[21] International Agency for Research on Cancer [IARC]. IARC Working Group. Asbestos. Lyon: IARC, 1972.

[22] International Agency for Research on Cancer [IARC]. IARC monographs on the valuation of carcinogenic risk of chemicals to man. Some inorganic and organometalic compounds. Lyon: IARC, 1973. pp. 17–47.

[23] Mossman BT. Mechanism of asbestos carcinogenesis and toxicity. The amphibole hypothesis revised. Br J Ind Med. 1993;50:673–676. PubMed PMID: 8398854.

[24] Piolatto G, Negri E, La Vecchia C, Pira E, Decarli A, Peto J. An update of cancer mortality among chrysotile asbestos miners in Balangero, Northern Italy. Br J Ind Med. 1990;47:810–814. PubMed PMID: 2176805.

[25] Dodič-Fikfak M, Kriebel D, Quinn MM, Eisen EA, Wegman DH. A case control study of lung cancer and exposure to chrysotile and amphibole at a Slovenian asbestos-cement plant. Ann Occup Hyg. 2007;51:261–268. PubMed PMID: 17351264.

[26] Dodič-Fikfak M., Šešok J. National guidelines for asbestos: final report (1.12.1998–31.10.1999). Ljubljana: Inštitut za varovanje zdravja RS, 1999.

[27] Brodkin CA, Rosenstock L. Asbestos and asbestos-related pleural disease. In: Rosenstock L, Cullen MR, Brodkin CA, Redlich CA editors. Textbook of clinical occupational and environmental medicine. 2nd ed. Philadelphia, Edinburgh, New York, St Louis, Sydney, Toronto: Elsevier Saunders, 2005. pp. 364–377.

[28] Rom WN. Asbestosis, pleural fibrosis, and lung cancer. In: Rom WN, Markowitz SB, editors. Environmental and occupational medicine. 4th ed. Philadelphia, Baltimore, New York, London, Buenos Aires, Hong Kong, Sydney, Tokyo: Wolters Kluwer, Lippincott Williams & Wilkins, 2007. pp. 298–316.

[29] Frumkin H, Berlin J. Asbestos exposure and gastrointestinal malignancy review and meta-analysis. Am J Ind Med. 1988;14:79–95. PubMed PMID: 3189361.

[30] Mossman BT, Bignon J, Corn M, Seaton A, Gee JB. Asbestos: scientific developments and implications for public policy. Science. 1990;247:294–301. PubMed PMID: 2153315.

[31] Selikoff IJ, Seidman H. Asbestos-associated deaths among insulation workers in the United States and Canada, 1967–1987. Ann N Y Acad Sci. 1991;643:1–14. PubMed PMID: 1809121.

[32] McDonald JC, Liddell FD, Dufresne A, McDonald AD. The 1891–1920 birth cohort of Quebec chrysotile miners and millers: mortality 1976–88. Br J Ind Med. 1993;50:1073–1081. PubMed PMID: 8280638.

[33] Giaroli C, Belli S, Bruno C, Candela S, Grignoli M, Minisci S, Poletti R, Ricco G, Vecchi G, Venturi G, Ziccardi A, Comba P. Mortality study of asbestos cement workers. Int Arch Occup Environ Health. 1994;66:7–11. PubMed PMID: 7927845.

[34] Hughes JM. Human evidence: lung cancer mortality risk from chrysotile exposure. Ann Occup Hyg. 1994;38:555–560. PubMed PMID: 7978978.

[35] Jakobsson K, Albin M, Hagmar L. Asbestos, cement, and cancer in the right part of the colon. Occup Environ Med. 1994;51:95–101. PubMed PMID: 8111470.

[36] Tarchi M, Orsi D, Comba P, De Santis M, Pirastu R, Battista G, Valani M. Cohort mortality study of rock salt workers in Italy. Am J Ind Med. 1994;25:251–256. PubMed PMID: 8147397.

[37] Straif K, Benbrahim-Tallaa L, Baan R, Grosse Y, Secretan B, El Ghissassi F, Bouvard V, Guha N, Freeman C, Galichet L, Cogliano V. WHO International Agency for Research on Cancer Monograph Working Group. A review of human carcinogens–part C: metals, arsenic, dusts, and fibres. Lancet Oncol. 2009;10:453–454. PubMed PMID: 19998521.

[38] Hughes JM, Weill H. Asbestosis as a precursor of asbestos-related lung cancer: results of a prospective mortality study. Br J Ind Med. 1991;48:229–233. PubMed PMID: 2025587.

[39] Karjalainen A, Pukkala E, Kauppinen T, Partanen T. Incidence of cancer among Finnish patients with asbestos-related pulmonary or pleural fibrosis. Cancer Causes Control. 1999;10:51–57. PubMed PMID: 10334642.

[40] Weiss W. Asbestosis: a marker for the increased risk of lung cancer among workers exposed to asbestos. Chest. 1999;115:536–549. PubMed PMID: 10027457.

[41] Reid A, de Klerk N, Ambrosini GL, Olsen N, Pang SC, Berry G, Musk AW. The effect of asbestosis on lung cancer risk beyond the dose related effect of asbestos alone. Occup Environ Med. 2005;62:885–889. PubMed PMID: 16299098.

[42] Peters GA, Peters BJ. Medical aspects. In: Peters GA, Peters BJ, editors. Sourcebook on asbestos diseases: medical, legal, and engineering aspects. New York, London: Gerland STPM Press, 1980. pp. B2–B6.

[43] Speizer FE, Balmes JR. Environmental lung disease. In: Fauci AS, Braunwald E, Kasper DL, Hauser SL, Longo DL, Jameson JL, Loscalzo J, editors. Harrison's principles of internal medicine. 17th ed. New York, Chicago, San Francisco, Lisbon, London, Madrid, Mexico City, New Delhi, San Juan, Seoul, Singapore, Sydney, Toronto: McGraw Hill, 2008. pp. 1612–1614.

[44] Holland JP, Smith DD. Asbestos. In: Greenberg MI, Hamilton RJ, Phillips SD, McCluskey GJ, editors. Occupational, industrial, and environmental toxicology. 2nd ed. Philadelphia: Mosby, An Affiliate of Elsevier Science, 2003. pp. 655–658.

[45] Garcia JG, Griffith DE, Cohen AB, Callahan KS. Alveolar macrophages from patients with asbestos exposure release increased levels of leukotriene B4. Am Rev Respir Dis. 1989;139:1494–1501. PubMed PMID: 254324.

[46] Kamp DW, Graceffa P, Pryor WA, Weitzman SA. The role of free radicals in asbestos-induced diseases. Free Radic Biol Med. 1992;12:293–315. PubMed PMID: 1577332.

[47] Ding M, Dong Z, Chen F, Pack D, Ma WY, Ye J, Shi X, Castranova V, Vallyathan V. Asbestos induces activator protein-1 transactivation in transgenic mice. Cancer Res. 1999;59:1884–1889. PubMed PMID: 10213496.

[48] Kinnula VL. Oxidant and antioxidant mechanisms of lung disease caused by asbestos fibres. Eur Respir J 1999;14:706–16. PubMed PMID: 10543297.

[49] Li J, Huang B, Shi X, Castranova V, Vallyathan V, Huang C. Involvement of hydrogen peroxide in asbestos-induced NFAT activation. Mol Cell Biochem. 2002;234–235:161–168. PubMed PMID: 12162429.

[50] Xu A, Zhou H, Yu DZ, Hei TK. Mechanisms of the genotoxicity of crocidolite asbestos in mammalian cells: implication from mutation patterns induced by reactive oxygen species. Environ Health Perspect. 2002;110:1003–1008. PubMed PMID: 12361925.

[51] Castranova V. Role of nitric oxide in the progression of pneumoconiosis. Biochemistry [Mosc]. 2004;69:32–37. PubMed PMID: 14972015.

[52] Castranova V. Signaling pathways controlling the production of inflammatory mediators in response to crystalline silica exposure: role of reactive oxygen/nitrogen species. Free Radic Biol Med. 2004;37:916–925. PubMed PMID: 15336307.

[53] Fattman CL, Chang LY, Termin TA, Petersen L, Enghild JJ, Oury TD. Enhanced bleomycin-induced pulmonary damage in mice lacking extracellular superoxide dismutase. Free Radic Biol Med. 2003;35:763–771. PubMed PMID: 14583340.

[54] Fattman CL, Tan RJ, Tobolewski JM, Oury TD. Increased sensitivity to asbestos-induced lung injury in mice lacking extracellular superoxide dismutase. Free Radic Biol Med. 2006;40:601–6007. PubMed PMID: 16458190.

[55] Kinnula VL, Hodgson UA, Lakari EK, Tan RJ, Sormunen RT, Soini YM, Kakko SJ, Laitinen TH, Oury TD, Pääkkö PK. Extracellular superoxide dismutase has a highly specific localization in idiopathic pulmonary fibrosis/usual interstitial pneumonia. Histopathology. 2006;49:66–74. PubMed PMID: 16842247.

[56] Weitzman SA, Graceffa P. Asbestos catalyzes hydroxyl and superoxide radical generation from hydrogen peroxide. Arch Biochem Biophys. 1984;228:373–376. PubMed PMID: 6320737.

[57] Halliwell B, Gutteridge JMC. Reactive species as useful biomolecules. In: Halliwell B, Gutteridge JMC, editors. Free radicals in biology and medicine. 3rd ed. Oxford, New York: Oxford University Press, 1999. p. 464.

[58] Kamp DW, Panduri V, Weitzman SA, Chandel N. Asbestos-induced alveolar epithelial cell apoptosis: role of mitochondrial dysfunction caused by iron-derived free radicals. Mol Cell Biochem. 2002;234–235:153–160. PubMed PMID: 12162428.

[59] Panduri V, Weitzman SA, Chandel N, Kamp DW. The mitochondria-regulated death pathway mediates asbestos-induced alveolar epithelial cell apoptosis. Am J Respir Cell Mol Biol. 2003;28:241–248. PubMed PMID: 12540492.

[60] Poser I, Rahman Q, Lohani M, Yadav S, Becker HH, Weiss DG, Schiffmann D, Dopp E. Modulation of genotoxic effects in asbestos-exposed primary human mesothelial cells by radical scavengers, metal chelators and a glutathione precursor. Mutat Res. 2004;559:19–27. PubMed PMID: 15066570.

[61] Thomas G, Ando T, Verma K, Kagan E. Asbestos fibers and interferon-gamma upregulate nitric oxide production in rat alveolar macrophages. Am J Respir Cell Mol Biol. 1994;11:707–715. PubMed PMID: 7524571.

[62] Quinlan TR, BeruBe KA, Hacker MP, Taatjes DJ, Timblin CR, Goldberg J, Kimberley P, O'Shaughnessy P, Hemenway D, Torino J, Jimenez LA, Mossman BT. Mechanisms of asbestos-induced nitric oxide production by rat alveolar macrophages in inhalation and in vitro models. Free Radic Biol Med. 1998;24:778–788. PubMed PMID: 9586808.

[63] Tanaka S, Choe N, Hemenway DR, Zhu S, Matalon S, Kagan E. Asbestos inhalation induces reactive nitrogen species and nitrotyrosine formation in the lungs and pleura of the rat. J Clin Invest. 1998;102:445–454. PubMed PMID: 9664087.

[64] Aldieri E, Ghigo D, Tomatis M, Prandi L, Fenoglio I, Costamagna C, Pescarmona G, Bosia A, Fubini B. Iron inhibits the nitric oxide synthesis elicited by asbestos in murine macrophages. Free Radic Biol Med. 2001;31:412–417. PubMed PMID: 11461780.

[65] Förstermann U, Kleinert H. Nitric oxide synthase: expression and expressional control of the three isoforms. Naunyn Schmiedebergs Arch Pharmacol. 1995;352:351–364. PubMed PMID: 8532063.

[66] deRojas-Walker T, Tamir S, Ji H, Wishnok JS, Tannenbaum SR. Nitric oxide induces oxidative damage in addition to deamination in macrophage DNA. Chem Res Toxicol. 1995;8:473–477. PubMed PMID: 7578935.

[67] Beckman JS, Koppenol WH. Nitric oxide, superoxide, and peroxynitrite: the good, the bad, and ugly. Am J Physiol. 1996;271:C1424–C1437. PubMed PMID: 8944624.

[68] Wink DA, Mitchell JB. Chemical biology of nitric oxide: insights into regulatory, cytotoxic, and cytoprotective mechanisms of nitric oxide. Free Radic Biol Med. 1998;25:434–456. PubMed PMID: 9741580.

[69] Halliwell B, Gutteridge JMC. The chemistry of free radicals and related "reactive species". In: Halliwell B, Gutteridge JMC, editors. Free radicals in biology and medicine. 3rd ed. Oxford, New York: Oxford University Press, 1999. pp. 73–82.

[70] Chao CC, Park SH, Aust AE. Participation of nitric oxide and iron in the oxidation of DNA in asbestos-treated human lung epithelial cells. Arch Biochem Biophys. 1996;326:152–157. PubMed PMID: 8579364.

[71] Park SH, Aust AE. Participation of iron and nitric oxide in the mutagenicity of asbestos in hgprt-, gpt+ Chinese hamster V79 cells. Cancer Res. 1998;58:1144–1148. PubMed PMID: 9515798.

[72] Dörger M, Allmeling AM, Kiefmann R, Schropp A, Krombach F. Dual role of inducible nitric oxide synthase in acute asbestos-induced lung injury. Free Radic Biol Med. 2002;33:491–501. PubMed PMID: 12160931.

[73] Halliwell B, Gutteridge JMC. Oxidative stress: adaptation, damage, repair and death. In: Halliwell B, Gutteridge JMC, editors. Free radicals in biology and medicine. 3rd ed. Oxford, New York: Oxford University Press, 1999. pp. 246–350.

[74] Halliwell B, Gutteridge JMC. Antioxidant defences. In: Halliwell B, Gutteridge JMC, editors. Free radicals in biology and medicine. 3rd ed. Oxford, New York: Oxford University Press, 1999. pp. 105–243.

[75] McCord JM, Fridovich I. Superoxide dismutase. An enzymic function for erythrocuprein (hemocuprein). J Biol Chem. 1969;244:6049–6055. PubMed PMID: 5389100.

[76] Quan F, Korneluk RG, Tropak MB, Gravel RA. Isolation and characterization of the human catalase gene. Nucleic Acids Res. 1986;14:5321–5335. PubMed PMID: 3755525.

[77] Hendrickson DJ, Fisher JH, Jones C, Ho YS. Regional localization of human extracellular superoxide dismutase gene to 4pter-q21. Genomics. 1990;8:736–738. PubMed PMID: 2276747.

[78] Church SL, Grant JW, Meese EU, Trent JM. Sublocalization of the gene encoding manganese superoxide dismutase [MnSOD/SOD2] to 6q25 by fluorescence in situ hybridization and somatic cell hybrid mapping. Genomics. 1992;14:823–825. PubMed PMID: 1427917.

[79] Folz RJ, Abushamaa AM, Suliman HB. Extracellular superoxide dismutase in the airways of transgenic mice reduces inflammation and attenuates lung toxicity following hyperoxia. J Clin Invest. 1999;103:1055–1066. PubMed PMID: 10194479.

[80] Forsberg L, Lyrenäs L, de Faire U, Morgenstern R. A common functional C-T substitution polymorphism in the promoter region of the human catalase gene influences transcription factor binding, reporter gene transcription and is correlated to blood catalase levels. Free Radic Biol Med. 2001;30:500–505. PubMed PMID: 11182520.

[81] Bowler RP, Nicks M, Warnick K, Crapo JD. Role of extracellular superoxide dismutase in bleomycin-induced pulmonary fibrosis. Am J Physiol Lung Cell Mol Physiol. 2002;282:L719-L726. PubMed PMID: 11880297.

[82] Zelko IN, Mariani TJ, Folz RJ. Superoxide dismutase multigene family: a comparison of the CuZn-SOD (SOD1), Mn-SOD (SOD2), and EC-SOD (SOD3) gene structures, evolution, and expression. Free Radic Biol Med. 2002;33:337–349. PubMed PMID: 12126755.

[83] Carlsson LM, Jonsson J, Edlund T, Marklund SL. Mice lacking extracellular superoxide dismutase are more sensitive to hyperoxia. Proc Natl Acad Sci U S A. 1995;92:6264–6268. PubMed PMID: 7603981.

[84] Singhal SS, Saxena M, Ahmad H, Awasthi S, Haque AK, Awasthi YC. Glutathione S-transferases of human lung: characterization and evaluation of the protective role of the alpha-class isozymes against lipid peroxidation. Arch Biochem Biophys. 1992;299:232–241. PubMed PMID: 1444461.

[85] Hubbard NE, Erickson KL. Role of 5′-lipoxygenase metabolites in the activation of peritoneal macrophages for tumoricidal function. Cell Immuno. 1995;160:115–122. PubMed PMID: 7842477.

[86] Kelsey KT, Nelson HH, Wiencke JK, Smith CM, Levin S. The glutathione S-transferase theta and mu deletion polymorphisms in asbestosis. Am J Ind Med. 1997;31:274–279. PubMed PMID: 9055949.

[87] Ketterer B. A bird's eye view of the glutathione transferase field. Chem Biol Interact. 2001;138:27–42. PubMed PMID: 11640913.

[88] Hayes JD, Pulford DJ. The glutathione S-transferase supergene family: regulation of GST and the contribution of the isoenzymes to cancer chemoprotection and drug resistance. Crit Rev Biochem Mol Biol. 1995;30:445–600. PubMed PMID: 8770536.

[89] Timbrell J. Phase 2 reactions. Conjugation. In: Timbrell J, editors. Principles of biochemical toxicology. London, Philadelphia: Taylor & Francis, 2003. pp. 95–101.

[90] Hayes JD, Strange RC. Glutathione S-transferase polymorphisms and their biological consequences. Pharmacology .2000;61:154–166. PubMed PMID: 10971201.

[91] Hayes JD, Flanagan JU, Jowsey IR. Glutathione transferases. Annu Rev Pharmacol Toxicol. 2005;45:51–88. PubMed PMID: 15822171.

[92] Blackburn AC, Woollatt E, Sutherland GR, Board PG. Characterization and chromosome location of the gene GSTZ1 encoding the human Zeta class glutathione transferase and maleylacetoacetate isomerase. Cytogenet Cell Genet. 1998;83:109–114. PubMed PMID: 9925947.

[93] Anttila S, Hirvonen A, Vainio H, Husgafvel-Pursiainen K, Hayes JD, Ketterer B. Immunohistochemical localization of glutathione S-transferases in human lung. Cancer Res. 1993;53:5643–5648. PubMed PMID: 8242618.

[94] Stücker I, Boffetta P, Antilla S, Benhamou S, Hirvonen A, London S, Taioli E. Lack of interaction between asbestos exposure and glutathione S-transferase M1 and T1 genotypes in lung carcinogenesis. Cancer Epidemiol Biomarkers Prev. 2001;10:1253–8. PubMed PMID: 11751442.

[95] Hirvonen A, Tuimala J, Ollikainen T, Linnainmaa K, Kinnula V. Manganese superoxide dismutase genotypes and asbestos-associated pulmonary disorders. Cancer Lett. 2002;178:71–74. PubMed PMID: 11849743.

[96] Shimoda-Matsubayashi S, Matsumine H, Kobayashi T, Nakagawa-Hattori Y, Shimizu Y, Mizuno Y. Structural dimorphism in the mitochondrial targeting sequence in the human manganese superoxide dismutase gene. A predictive evidence for conformational change to influence mitochondrial transport and a study of allelic association in Parkinson's disease. Biochem Biophys Res Commun. 1996;226:561–565. PubMed PMID: 8806673.

[97] Rosenblum JS, Gilula NB, Lerner RA. On signal sequence polymorphisms and diseases of distribution. Proc Natl Acad Sci U S A. 1996;93:4471–3. PubMed PMID: 8633092.

[98] Ali-Osman F, Akande O, Antoun G, Mao JX, Buolamwini J. Molecular cloning, characterization, and expression in *Escherichia coli* of full-length cDNAs of three human

glutathione S-transferase P1 gene variants – evidence for differential conjugation capacity of the encoded proteins. J Biol Chem. 1997;272:10004–10012. PubMed PMID: 9092542.

[99] Tatemichi M, Sawa T, Gilibert I, Tazawa H, Katoh T, Ohshima H. Increased risk of intestinal type of gastric adenocarcinoma in Japanese women associated with long forms of CCTTT pentanucleotide repeat in the inducible nitric oxide synthase promoter. Cancer Lett. 2005;217:197–202. PubMed PMID: 15617837.

[100] Kinnula VL, Lehtonen S, Koistinen P, Kakko S, Savolainen M, Kere J, Ollikainen V, Laitinen T. Two functional variants of the superoxide dismutase genes in Finnish families with asthma. Thorax. 2004;59:116–119. PubMed PMID: 14760150.

[101] Gao F, Kinnula VL, Myllärniemi M, Oury TD. Extracellular superoxide dismutase in pulmonary fibrosis. Antioxid Redox Signal. 2008;10:343–354. PubMed PMID: 17999630.

[102] Kliment CR, Tobolewski JM, Manni ML, Tan RJ, Enghild J, Oury TD. Extracellular superoxide dismutase protects against matrix degradation of heparan sulfate in the lung. Antioxid Redox Signal. 2008;10:261–268. PubMed PMID: 17961072.

[103] Folz RJ, Peno-Green L, Crapo JD. Identification of a homozygous missense mutation [Arg to Gly] in the critical binding region of the human EC-SOD gene [SOD3] and its association with dramatically increased serum enzyme levels. Hum Mol Genet. 1994;3:2251–2254. PubMed PMID: 7881430.

[104] Sandström J, Nilsson P, Karlsson K, Marklund SL. 10-fold increase in human plasma extracellular superoxide dismutase content caused by a mutation in heparin-binding domain. J Biol Chem. 1994;269:19163–19166. PubMed PMID: 8034674.

[105] Yamada H, Yamada Y, Adachi T, Goto H, Ogasawara N, Futenma A, Kitano M, Hirano K, Kato K. Molecular analysis of extracellular-superoxide dismutase gene associated with high level in serum. Jpn J Hum Genet. 1995;40:177–184. PubMed PMID: 7662997.

[106] Jiang Z, Akey JM, Shi J, Xiong M, Wang Y, Shen Y, Xu X, Chen H, Wu H, Xiao J, Lu D, Huang W, Jin L. A polymorphism in the promoter region of catalase is associated with blood pressure levels. Hum Genet. 2001;109:95–98. PubMed PMID: 11479740.

[107] Casp CB, She JX, McCormack WT. Genetic association of the catalase gene [CAT] with vitiligo susceptibility. Pigment Cell Res. 2002;15:62–66. PubMed PMID: 11837458.

[108] Nadif R, Mintz M, Jedlicka A, Bertrand JP, Kleeberger SR, Kauffmann F. Association of CAT polymorphisms with catalase activity and exposure to environmental oxidative stimuli. Free Radic Res. 2005;39:1345–1350. PubMed PMID: 16298864.

[109] Zhou XF, Cui J, DeStefano AL, Chazaro I, Farrer LA, Manolis AJ, Gavras H, Baldwin CT. Polymorphisms in the promoter region of catalase gene and essential hypertension. Dis Markers. 2005;21:3–7. PubMed PMID: 15735318.

[110] Ahn J, Nowell S, McCann SE, Yu J, Carter L, Lang NP, Kadlubar FF, Ratnasinghe LD, Ambrosone CB. Associations between catalase phenotype and genotype: modification

by epidemiologic factors. Cancer Epidemiol Biomarkers Prev. 2006;15:1217–1222. PubMed PMID: 16775184.

[111] Bastaki M, Huen K, Manzanillo P, Chande N, Chen C, Balmes JR, Tager IB, Holland N. Genotype-activity relationship for Mn-superoxide dismutase, glutathione peroxidase 1 and catalase in humans. Pharmacogenet Genom. 2006;16:279–286. PubMed PMID: 16538174.

[112] Johansson AS, Stenberg G, Widersten M, Mannervik B. Structure-activity relationships and thermal stability of human glutathione transferase P1-1 governed by the H-site residue 105. J Mol Bio. 1998;278:687–698. PubMed PMID: 9600848.

[113] Watson MA, Stewart RK, Smith GB, Massey TE, Bell DA. Human glutathione S-transferase P1 polymorphisms: relationship to lung tissue enzyme activity and population frequency distribution. Carcinogenesis. 1998;19:275–280. PubMed PMID: 9498276.

[114] Xu W, Liu L, Emson PC, Harrington CR, Charles IG. Evolution of a homopurine-homopyrimidine pentanucleotide repeat sequence upstream of the human inducible nitric oxide synthase gene. Gene. 1997;204:165–170. PubMed PMID: 9434180.

[115] Warpeha KM, Xu W, Liu L, Charles IG, Patterson CC, Ah-Fat F, Harding S, Hart PM, Chakravarthy U, Hughes AE. Genotyping and functional analysis of a polymorphic [CCTTT][n] repeat of NOS2A in diabetic retinopathy. FASEB J. 1999;13:1825–1832. PubMed PMID: 10506586.

[116] Franko A, Dolžan V, Arnerić N, Dodič-Fikfak M. The influence of genetic polymorphisms of GSTP1 on the development of asbestosis. J Occup Environ Med. 2008;50:7–12. PubMed PMID: 18188076; doi: 10.1097/JOM.0b013e31815cbab5.

[117] Franko A, Dolžan V, Arnerić N, Dodič-Fikfak M. Asbestosis and catalase genetic polymorphism. Arh Hig Rada Toksikol. 2008;59:233–240. PubMed PMID: 19064360; doi: 10.2478/10004-1254-59-2008-1907.

[118] Franko A, Dodič-Fikfak M, Arnerić N, Dolžan V. Manganese and extracellular superoxide dismutase polymorphisms and risk for asbestosis. J Biomed Biotechnol. 2009;2009:493083. PubMed PMID: 19636420; doi: 10.1155/2009/493083.

[119] Franko A, Dodič-Fikfak M, Arnerić N, Dolžan V. Inducible nitric oxide synthase genetic polymorphism and risk of asbestosis. J Biomed Biotechnol. 2011;2011:685870. PubMed PMID: 21660141; doi: 10.1155/2011/685870.

[120] Ferris BG. Epidemiology standardization project [American Thoracic Society]. Am Rev Respir Dis. 1978;118:1–120. PubMed PMID: 742764.

[121] Asbestos, asbestosis, and cancer: the Helsinki criteria for diagnosis and attribution. Consensus report. Scand J Work Environ Health. 1997;23:311–316. PubMed PMID: 9322824.

[122] American Thoracic Society. Diagnosis and initial management of nonmalignant diseases related to asbestos. Am J Respir Crit Care Med. 2004;170:691–751. PubMed PMID: 15355871.

[123] Yoshie Y, Ohshima H. Nitric oxide synergistically enhances DNA strand breakage induced by polyhydroxyaromatic compounds, but inhibits that induced by the Fenton reaction. Arch Biochem Biophys. 1997;342:13–21. PubMed PMID: 9185609.

[124] Haberstroh K, Heigold S, Bauer G. Transformed cell-derived reactive oxygen species support and inhibit nitric oxide-mediated apoptosis induction. Int J Oncol. 2002;21:145–151. PubMed PMID: 12063561.

[125] Jackson JH, Schraufstatter IU, Hyslop PA, Vosbeck K, Sauerheber R, Weitzman SA, Cochrane CG. Role of oxidants in DNA damage. Hydroxyl radical mediates the synergistic DNA damaging effects of asbestos and cigarette smoke. J Clin Invest. 1987;80:1090–1095. PubMed PMID: 2821073.

[126] Valavanidis A, Vlachogianni T, Fiotakis K. Tobacco smoke: involvement of reactive oxygen species and stable free radicals in mechanisms of oxidative damage, carcinogenesis and synergistic effects with other respirable particles. Int J Environ Res Public Health. 2009;6:445–462. PubMed PMID: 19440393; doi: 10.3390/ijerph6020445.

[127] Valavanidis A, Balomenou H, Macropoulou I, Zarodimos I. A study of the synergistic interaction of asbestos fibers with cigarette tar extracts for the generation of hydroxyl radicals in aqueous buffer solution. Free Radic Biol Med. 1996;20:853–858. PubMed PMID: 8728034.

[128] van der Vliet A, Cross CE. Oxidants, nitrosants, and the lung. Am J Med. 2000;109:398–421. PubMed PMID: 11020397.

[129] Hasnis E, Bar-Shai M, Burbea Z, Reznick AZ. Mechanisms underlying cigarette smoke-induced NF-kappaB activation in human lymphocytes: the role of reactive nitrogen species. J Physiol Pharmacol. 2007;58:275–287. PubMed PMID: 18204137.

[130] Figueroa JD, Koutros S, Colt JS, Kogevinas M, Garcia-Closas M, Real FX, Friesen MC, Baris D, Stewart P, Schwenn M, Johnson A, Karagas MR, Armenti KR, Moore LE, Schned A, Lenz P, Prokunina-Olsson L, Banday AR, Paquin A, Ylaya K, Chung JY, Hewitt SM, Nickerson ML, Tardón A, Serra C, Carrato A, García-Closas R, Lloreta J, Malats N, Fraumeni JF Jr, Chanock SJ, Chatterjee N, Rothman N, Silverman DT. Modification of Occupational Exposures on Bladder Cancer Risk by Common Genetic Polymorphisms. J Natl Cancer Inst. 2015; 107(11). pii: djv223. PubMed PMID: 26374428; doi: 10.1093/jnci/djv223.

[131] Huo X, Chen D, He Y, Zhu W, Zhou W, Zhang J. Bisphenol-A and female infertility: a possible role of gene-environment interactions. Int J Environ Res Public Health. 2015;12:11101–11116. PubMed PMID: 26371021; doi: 10.3390/ijerph120911101.

[132] Melkonian SC, Daniel CR, Ye Y, Tannir NM, Karam JA, Matin SF, Wood CG, Wu X.Gene-environment interaction of genome-wide association study-identified susceptibility

loci and meat-cooking mutagens in the etiology of renal cell carcinoma. Cancer. 2016;122:108–115. PubMed PMID: 26551148; doi: 10.1002/cncr.29543.

[133] Reynolds CA, Gatz M, Christensen K, Christiansen L, Dahl Aslan AK, Kaprio J, Korhonen T, Kremen WS, Krueger R, McGue M, Neiderhiser JM, Pedersen NL, IGEMS consortium. Gene-environment interplay in physical, psychological, and cognitive domains in mid to late adulthood: is APOE a variability gene? Behav Genet. 2016;46:4–19. PubMed PMID: 26538244; doi: 10.1007/s10519-015-9761-3.

[134] Tunesi S, Ferrante D, Mirabelli D, Andorno S, Betti M, Fiorito G, Guarrera S, Casalone E, Neri M, Ugolini D, Bonassi S, Matullo G, Dianzani I, Magnani C. Gene-asbestos interaction in malignant pleural mesothelioma susceptibility. Carcinogenesis. 2015;36:1129–1135. PubMed PMID: 26139392; doi: 10.1136/oemed-2015-102803.

[135] Malhotra J, Sartori S, Brennan P, Zaridze D, Szeszenia-Dabrowska N, Świątkowska B, Rudnai P, Lissowska J, Fabianova E, Mates D, Bencko V, Gaborieau V, Stücker I, Foretova L, Janout V, Boffetta P. Effect of occupational exposures on lung cancer susceptibility: a study of gene-environment interaction analysis. Cancer Epidemiol Biomarkers Prev. 2015;24:570–579. PubMed PMID: 25583949; doi: 10.1158/1055-9965.

[136] Liu CY, Stücker I, Chen C, Goodman G, McHugh MK, D'Amelio AM Jr, Etzel CJ, Li S, Lin X, Christiani DC. Genome-wide gene-asbestos exposure interaction association study identifies a common susceptibility variant on 22q13.31 associated with lung cancer risk. Cancer Epidemiol Biomarkers Prev. 2015;24:1564–1573. PubMed PMID: 26199339; doi: 10.1158/1055-9965.EPI-15-0021.

[137] Schulte PA, Whittaker C, Curran CP. Considerations for using genetic and epigenetic information in occupational health risk assessment and standard setting. J Occup Environ Hyg. 2015;12 Suppl 1:S69-S81. PubMed PMID: 26583908; doi: 10.1080/15459624.2015.1060323.

Section 3

Work-Related Musculoskeletal Disorders

Chapter 9

Ibuprofen as a Treatment for Work-Related Musculoskeletal Disorders: Effectiveness versus Caveats

Mary F. Barbe and Ann E. Barr-Gillespsie

Additional information is available at the end of the chapter

http://dx.doi.org/10.5772/66480

Abstract

Work-related upper limb disorders (WMSDs), also known as repetitive strain injuries, affect a large subsection of the US population. These disorders are a significant source of injury, morbidity, loss of work, and pain. We have developed a rat model of upper extremity repetitive work at high forces, and observed exposure-dependent increased inflammatory responses in all tissues involved in performing the task. A 2- to 8-week regimen of oral ibuprofen provided to rats while they continued to perform a high-repetition high-force task ameliorated these inflammatory responses as well as several motor declines. Ibuprofen treatment also attenuated task-induced tissue fibrosis, cartilage degeneration, and bone osteopenia, indicating their link to inflammatory processes. However, ibuprofen did not significantly attenuate persistent nocifensive pain behaviors (reflexive grip strength results are presented) likely because of persistent increases in inflammatory cytokines in the spinal cord, suggestive of central sensitization. Since long-term ibuprofen use can induce a number of negative side effects, such as gastritis, multi-pronged approaches should be considered with anti-inflammatory drugs included for only short time periods.

Keywords: repetitive loading, work-related musculoskeletal disorders, repetitive strain injury, ibuprofen, osteopenia

1. Introduction

Overuse-induced musculoskeletal disorders (MSDs) are also known as overuse injuries, repetitive strain injuries. Diagnoses of upper extremity MSDs include muscle strain injuries, carpal and cubital tunnel syndromes, muscle myalgia/hyperalgesia, dorsal wrist tendinosis, lateral and medial epicondylopathies, rotator cuff tendinopathies, and more. These disorders often occur as a consequence of daily activities (both occupational and not), sports or military activities, and are a leading cause of pain and physical disability [1–4]. Some cases become so severe that simple personal tasks, such as buttoning a shirt, become difficult to impossible. Acute trauma may be a causal factor in some WMSDs. Yet, many result from cumulative small amplitude forces occurring with overtraining, overexertion, repetitive activities, forceful actions, and prolonged static positioning [5–8]. Prevention is hampered by many problems [9, 10]. There remains a call for effective treatments for these often debilitating disorders [9, 11].

2. Current treatments for overuse—MSDs

The first line of treatment for workers in pain usually entails a prescription of non-steroidal anti-inflammatory drugs (NSAIDs), such as ibuprofen [12, 13]. NSAIDs are the most commonly used (self-care and prescribed) for acute and chronic musculoskeletal pain [14–17]. A survey study of 941 workers found that 84% used NSAIDs, including ibuprofen, for pain [17]. Forty percent of 2213 French workers reported the regular use of ibuprofen in a 1-month period [18]. Back and shoulder injuries and other musculoskeletal strains are largely self-treated by migrant farm workers with rest and over-the-counter drugs, such as ibuprofen [14]. Rest, ice, compression, and elevation (termed RICE oftewn) are also often used to treat acute injuries. However, RICE has proved less effective for treatment of pain associated with chronic overuse—MSDs than NSAIDs. Splinting for carpal tunnel syndrome is less effective than surgical release or injections of steroids around the nerve [19–24], which are also not always effective [19–24].

3. Ibuprofen

Ibuprofen was introduced to the US market as a prescription drug to treat arthritic conditions in 1974, and subsequently became available over the counter in the United States in 1984. Despite its relatively short history as an over-the-counter medicine, it has quickly achieved popularity as a treatment for musculoskeletal and peripheral nerve pain, capturing up to a third of the over-the-counter analgesics market of the US by 2002. According to background information supplied by Wyeth Pharmaceuticals (a manufacturer of Advil, a brand name ibuprofen), this occurred principally because of its strong gastrointestinal (GI) safety profile that ibuprofen was approved for over-the-counter use. There are a wide variety of ibuprofen drugs available on the market as indicated in **Table 1**. Tablet, caplet, injectable, and topical forms are available. Negative side effects and major concerns will be discussed later in this chapter, although it should be noted that topical ibuprofen formulas are absorbed less into blood stream than oral forms, avoiding several side effects. However, as topical NSAID drugs are not systemic, they will not reduce inflammatory responses other than at the site of application.

Ibuprofen works by inhibiting both the constitutive cyclooxygenase (COX)-1 and the more inducible COX-2 enzyme. These enzymes catalyze the generation of prostanoids (prostaglandins PGE2 and PGF2a), prostacyclins, and thromboxanes [25, 26]. Inhibition of these enzymes by ibuprofen prevents the conversion of arachidonic acid to prostaglandin H2, and in doing so blocks the prostaglandin-signaling pathway. Prostaglandins play an important role in pain and inflammatory signaling, as well as have roles in maintaining kidney function (mainly by regulating blood flow in the glomerular capsule) and the gut mucosa, and cardiovascular physiological processes [26].

Brand names
Ibuprofen Tablets and Caplets: Actiprofen Caplets (CA), Advil, Advil Extra Strength (CA), Advil Migraine, Anadin Ibuprofen (UK), Anadin Ultra (UK), Apo-Ibuprofen (CA), Arthrofen (UK), Brufen (UK), Cuprofen (UK), Extra Strength Motrin IB (CA), Hedex Ibuprofen (UK), Motrin, Motrin IB, Ibuprofen (CA), IBU
Active ingredient: Ibuprofen (100–800-mg tables and caplets available). **Typical dose** is 200–400 mg/dose; Maximum amount is 800 mg/dose, or 3200 mg per day. **Use:** Reduction of fever, pain, or inflammation from headache, dental pain, menstrual cramps, rheumatoid arthritis, osteoarthritis, muscle aches, minor aches, and pain. **Note:** An anti-inflammatory dose is higher than an analgesic dose, and must be maintained for full effectiveness.
Ibuprofen PM Tablets: Advil PM, Motrin PM
Active Ingredient: Ibuprofen (200 mg) and Diphenhydramine citrate (38 mg). **Typical dose** is two capsules at bedtime (also the maximum dose/day). **Use:** Occasional sleeplessness when associated with minor aches and pains.
Injectable Ibuprofen: Caldolor, Calprofen (UK), and more
Active Ingredient: Ibuprofen (various doses available). **Typical dose**: Intravenous infusion of 100–800 mg dose, after dilution to 4 mg/ml or less per injection. **Use:** Reduction of fever; Management of mild to moderate pain, and moderate to severe pain as an adjunct to opioid analgesics.
Topical Ibuprofen: Ibuprofen Gel (US), Ibuleve gel (UK), Ibumousse (UK), Ibuspray (UK), and more
Active Ingredient: Ibuprofen (various doses available). **Typical dose:** three to four times a day, or as directed by a doctor, with at least 4 h between applications. **Use:** Muscle or rheumatic pain, backache, neuralgia; sprains, strains and sports injuries; mild arthritis. **Note:** Absorbed less into blood stream when applied topically, so not thought to reduce fever or widespread inflammation as a consequence.

Table 1. Types of ibuprofen available.

A steady dose of ibuprofen is considered necessary to attenuate the increase in inflammation, rather than just analgesic. The dose used should be lower than the maximum limit for gastrointestinal toxicity. Those suggested maximum limits are indicated in **Table 1**.

4. An operant rat model of WMSD

Several animal models have been developed to study WMSDs and have shown that repetitive hand activities induce sensorimotor dysfunction [27–33]. A model developed in our laboratory is a unique operant rat model of voluntary reaching and grasping (**Figure 1**; [7, 34]). Using this model, we are able to examine the effects of voluntary performance of repetitive low or high demand tasks on sensorimotor performance and musculoskeletal tissues [7, 30, 35]. This model is nonsurgical and involves performance of voluntary repetitive tasks to induce mechanical loading of forearm tissues. Specifically, adult rats are required to voluntarily and repetitively reach for, grasp, and isometrically pull a handle with one forelimb to obtain a food reward at various reach rates and force levels determined from studies on risk exposure for WMSDs to humans [7, 34]. Additionally, several functional outcomes are tested that are similar to those tested in patients, including forepaw (hand) sensitivity, grip strength, and median nerve conduction velocity.

Using this rat model, we have observed early exposure-dependent changes (duration and task level) in inflammatory responses in the form of increased macrophages and inflammatory cytokines in soft tissues involved in performing the repetitive task [7, 30, 32, 35, 36]. The greatest responses were observed in rats performing a high-repetition high-force (HRHF) task for 6–12 weeks, compared to lower demand tasks. Therefore, we picked this HRHF task regimen for experiments in which we tested the effectiveness of ibuprofen.

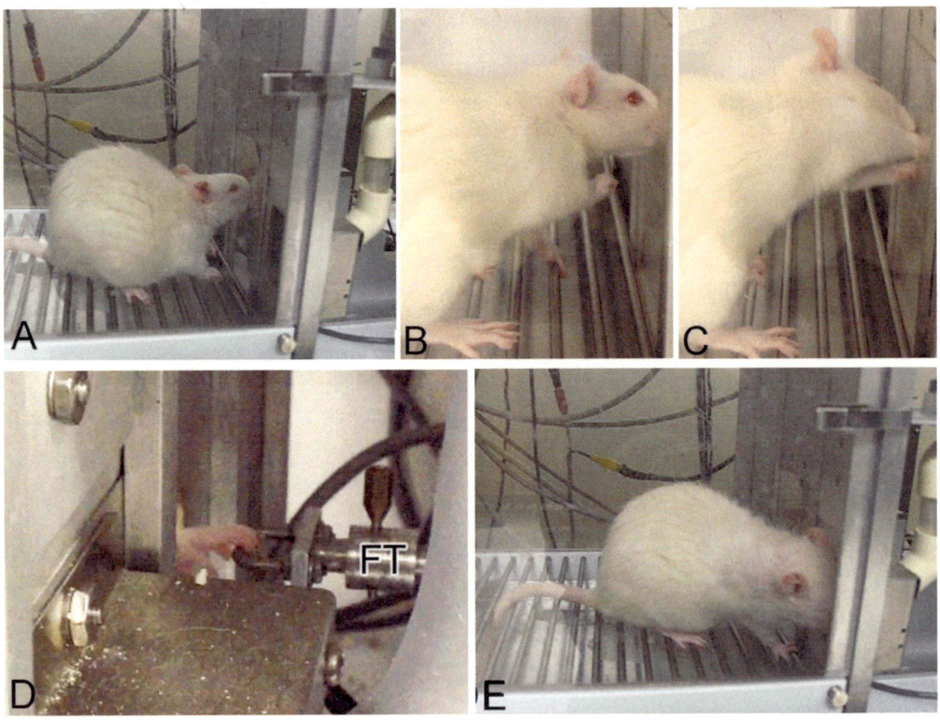

Figure 1. Rat performing HRHF repetitive reaching task. (A) Rat awaits auditory stimulus with snout in portal. (B and C) Rat reaches for force handle with right forepaw; left forepaw used for postural support. (D) Closer view, rat grasps and isometrically pulls force handle attached to force transducer (FT), until predetermined force threshold is reached and held for at least 50 ms. (E) Rat retrieves foot pellet reward by mouth from food trough.

http://dx.doi.org/10.5772/66480

5. Testing the effectiveness of ibuprofen treatment for WMSDs

We hypothesized that an underlying inflammatory mechanism is driving many of the sensorimotor declines, as are inflammation-linked fibrotic and degenerative/degradative tissue changes [37]. We explored this hypothesis by treating rats with systemic ibuprofen (i.e., oral) at anti-inflammatory doses. The design of these experiments is shown in **Figure 2**, and included normal controls (termed NC rats) and food-restricted-only controls (termed FRC rats). Rats were food-restricted to body weights of 5% less than age-matched normal controls to motivate them to work. Subsets of food-restricted rats were trained to high-force levels to determine the effects of training (10 min/day, 5 days/week, for 5 weeks) in which they learn to pull at high-force levels (1.25 Newton's which is approximately 60% of their maximum voluntary force) [7]. The trained-only rats are termed TRHF rats. Subsets of TRHF rats went on to perform a high-repetition high-force task regimen for 2 h/day, 3 days/week for up to 12 weeks. Task requirements were a reach rate of 8 reaches/min and a target force of 60 ± 5% of their mean maximum pulling force. HRHF rats had to grasp the force lever bar and exert an isometric pull at the target level for at least 50 ms to receive a food reward. Half of each group was administered ibuprofen (Children's Motrin Grape Flavored, Johnson & Johnson) in drinking water daily (a dose of 45 mg/kg body weight was used). This dose was lower than the maximum limit for gastrointestinal toxicity in rats, yet effective in reducing chronic inflammation [38]. The results of these experiments and the effectiveness (or lack thereof) are discussed below.

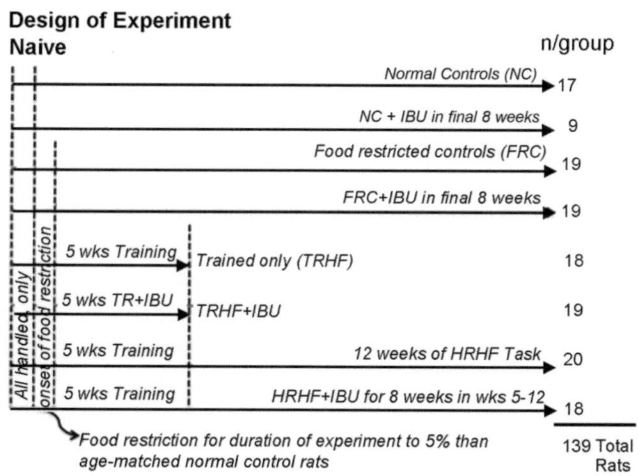

Figure 2. Experimental design. (A) Food restriction began after a 1-week period of daily handling. All rats but normal control (NC ± ibuprofen treatment) were food restricted to 5% less than weights of age-matched NC rats. NC and food-restricted control (FRC) rats rested until euthanasia at matched time points as HRHF rats. Trained and task rats underwent a 5-week training period (rats reached the HRHF level by last week of training). These trained-only rats (TRHF) were euthanized after training. Task rats performed a high-repetition high-force (HRHF) task for 12 weeks. NC+IBU and FRC+IBU rats received daily ibuprofen (IBU) treatment of 45 mg/kg of body weight in drinking water in the final 8 weeks, as did HRHF+IBU rats (arrow indicates the onset of ibuprofen treatment). TRHF+IBU rats received ibuprofen treatment prophylactically during training. The number of rats per group is shown at the far right.

5.1. Ibuprofen effectively reduced tissue inflammation induced by the HRHF task and voluntary motor abilities

Some mechanisms examined to date in our rat model include task-induced tissue injury, inflammation, and fibrosis, each of which contributed to declines in grip strength by producing discomfort or affecting biomechanical strength. Evidence of tissue injury was paralleled by inflammatory responses, such as increased pro-inflammatory cytokines in flexor digitorum muscles and tendons [30–32, 39], and increased macrophages in the median nerve at the level of the wrist (**Figure 3A, B**). Elevated levels of key pro-inflammatory cytokines, IL-1beta, and TNF-alpha were also observed in serum of rats that had performed a HRHF task for 12 weeks (12-week HRHF rats) (**Figure 3D, E**).

Treatment of rats performing a high-repetition high-force task with oral ibuprofen in weeks 5–12 of a 12-week task regimen significantly reduced macrophage numbers and inflammatory cytokines in tissues and serum (**Figure 3A, B, D**). Ibuprofen treatment also improved HRHF-induced declines in several voluntary work parameters, including reach rate, voluntary pulling force, and duration of voluntary performance (**Figure 4A, B**) [40, 41]. Similarly, the treatment of human subjects with ibuprofen before unaccustomed exercise improves muscle strength [42, 43]. The attenuation of voluntary reach abilities in HRHF+IBU rats in parallel with reduced numbers of macrophages in the median nerve (**Figure 3A**) [40] indicates that a task-induced neuralgia is contributing to voluntary motor declines seen in **Figure 4A and B**.

5.2. Ibuprofen treatment did not ameliorate HRHF-induced spinal cord sensitization or muscle hyperalgesia

However, reflexive grip strength was not rescued in 9- and 12-week HRHF+IBU rats (**Figure 4C**) [41]. This type of nocifensive motor behavior has been termed muscle hyperalgesia [44] and is a type of chronic pain. We postulate that ibuprofen did not rescue reflexive grip strength declines because it did not prevent inflammation-associated changes in the central nervous system. We stained cervical regions of the spinal cord for pro-inflammatory cytokine IL-1-beta levels using immunohistochemical methods and found that both untreated HRHF and HRHF+IBU animals expressed this cytokine in neurons and some glial cells at roughly the same frequency and intensity (**Figure 5A, B, D**). This was in sharp contrast to IL-1-beta immunoexpression in spinal cords of normal control rats, which showed an almost absence of IL-1-beta immunoexpression (**Figure 5C**). We postulate that ibuprofen, or other anti-inflammatory drug, would have to be provided earlier than week 4 prior to the onset of pain behaviors in order to be fully effective. Future studies need to consider these negative central nervous system changes to successfully treat chronic pain behaviors in subjects with WMSDs.

5.3. HRHF task-induced tissue fibrosis is effectively reduced by ibuprofen, indicative of an underlying inflammatory mechanism

Muscles undergo repetitive strain-induced fibrosis. Stauber and colleagues have shown that repeated muscle strains at fast velocities resulted in fibrotic myopathy with increased collagen content, collagen cross-links, and non-contractile tissues [45–48]. Factors and mechanisms of repetitive strain-induced fibrosis are still under investigation. They appear to involve transform-

http://dx.doi.org/10.5772/66480

ing growth factor beta-1 (TGFB-1) and connective tissue growth factor (CTGF), a key downstream mediator of TGFB-1's effects on matrix protein production [49–53]. Strong links between mechanical loading and increased TGFB-1 and CTGF protein levels in muscles and tendons *in vivo*, and in isolated fibroblasts and tenocytes, have been established [50, 52–55]. It is key to identify effective early or preventive treatments for such tissue fibrosis, as recovery from such tissue fibrosis is slow, even with complete cessation of strain or activity for up to 3 months [47].

CTGF production also appears to be regulated by pro-inflammatory cytokines, IL-1-beta, and TNF-alpha, which are also thought of as pro-fibrogenic cytokines [37, 56, 57]. Since we have observed that task-induced tissue inflammation precedes tissue fibrotic responses, including increased CTGF and collagen type 1 production [58–60], we next examined the effects of secondary ibuprofen treatment on fibrogenic processes in our rat model [40, 61]. In addition to successful reductions of tissue and serum inflammatory responses after ibuprofen treatment, we observed significant reductions in TGFB-1 and CTGF protein expression as well as collagen deposition in median nerves (**Figure 3A**) and flexor digitorum muscles of 6-week and 12-week HRHF+IBU rats (**Figure 6**) [40, 61]. These findings support an underlying inflammatory drive on at least some fibrogenic processes. This reduction in collagen deposition within and around tissue components of the upper extremity may also aid the return of function, such as the return of median nerve conduction velocity in median nerves of 12-week HRHF rats as shown in **Figure 3C**.

5.4. HRHF task-induced radiocarpal joint damage is ameliorated by ibuprofen treatment

Joint degeneration may occur for a number of reasons including joint trauma from increased repetition of joint loading, high impact joint loading, increased inflammatory processes (e.g., autoimmune), or pathological metabolic processes [62–64]. Radiocarpal and intercarpal joints of the wrist and hand, respectively, can show signs of increased incidence of hand osteoarthritis in individuals involved in intense (defined as long duration, high repetition, and/or high force) occupationally related physical activities [65–67]. A high incidence of radiographic of hand osteoarthritis has been identified in middle-aged female dentists and teachers [66, 67]. Several studies report that increasing radiographic severity of hand osteoarthritis is associated with reduced hand function and increased pain [66, 68, 69]. Therefore, the impact of hand osteoarthritis is considerable [68, 69].

After 12 weeks of performing the HRHF task, untreated task animals demonstrated evidence of joint inflammation (loss of proteoglycan staining as shown in **Figure 7B** as compared to controls in **Figure 7A**) [70]. This loss of proteoglycan staining in untreated 12-week HRHF rats is captured in the form of elevated Mankin histopathological scores (**Figure 7E**), a scoring system that also reflects a development of pannus and apoptotic cells in the joint cartilage. Each of these changes was indicative of task-induced joint degeneration. Serum biomarker testing revealed increased levels of a serum biomarker of collagen degradation, C1,2C (a marker of collagen type 1 and 2 degradation fragments produced by collagenase cleavage of type II collagen) in untreated 12-week HRHF rats [70]. Increased activated macrophages, cyclooxygenase immunopositive cells, and inflammatory cytokine levels were detected in the distal radius, ulna, and carpal bones (the latter shown in **Figure 7F, G**), supporting our hypothesis of an underlying inflammatory mechanism.

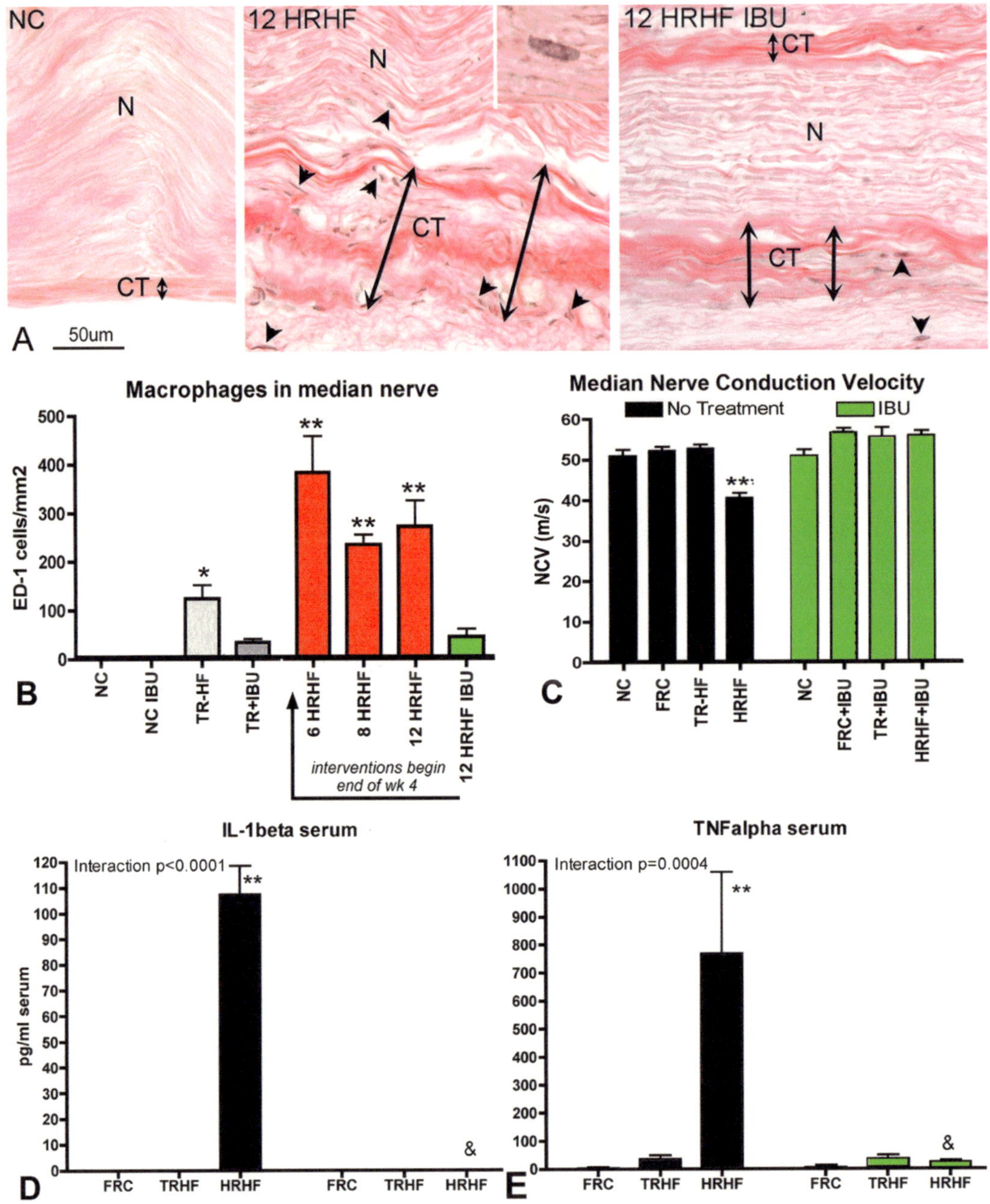

Figure 3. Median nerve inflammatory and fibrotic responses as well as systemic cytokine responses. (A) Photomicrographs showing increased activated macrophages in the median nerve of HRHF rats (detected immunohistochemically and denoted with arrowheads), and width of epineurial connective tissues (CT; (double arrows)) around the median nerve (N) at the wrist level. Eosin counterstain. (B) Mean number of activated macrophage in the median nerve decreased with ibuprofen treatment provided daily in task weeks 5–12. (C) Nerve conduction velocity (NCV) in meters/second (m/sec) declined in HRHF rats and was rescued by ibuprofen treatment that began after task week 4 (arrow) and that continued though task week 12. (D and E) IL-1-beta and TNF-alpha increased systemically (in serum) in untreated HRHF rats. These increases were ameliorated with 8 weeks of ibuprofen treatment. Symbols: $^{*}p < 0.05$ and $^{**}p < 0.01$, compared to NC or FRC rats; $^{\&}p < 0.05$, compared to untreated HRHF rats. Modified with permission from Jain et al. [40], and used by permission.

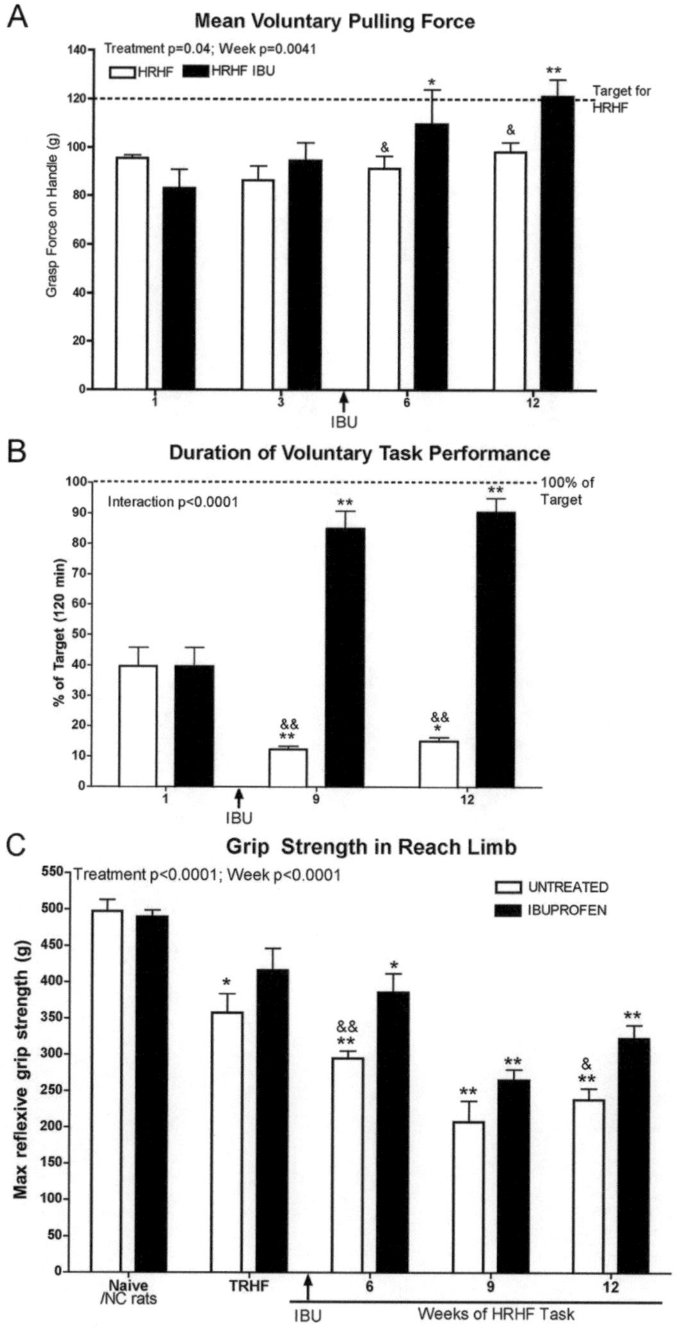

Figure 4. Voluntary and reflexive motor abilities. (A) Mean voluntary pulling force on the handle (percent of maximum pulling force) in grams. Across weeks of task performance, the mean voluntary pulling force was lower than target levels in untreated HRHF rats, yet met target levels in ibuprofen-treated rats (ibuprofen was provided in task weeks 5–12, with onset indicated by arrow). (B) Across the weeks, the duration of voluntary task performance decreased in untreated HRHF rats. By contrast, the duration was near target levels in HRHF+IBU rats in weeks 9 and 12. (C) Grip strength (maximum reflexive grip strength in grams) in the preferred reach limb decreased in both groups, compared to baseline naïve levels. Ibuprofen treatment only partially rescued this nocifensive motor behavior. $^{*}p < 0.05$ and $^{**}p < 0.01$, compared to week 1; $^{\&}p < 0.05$ and $^{\&}p < 0.01$, compared to target levels. Used by permission from Jain et al. [40] and Kietrys et al. [41].

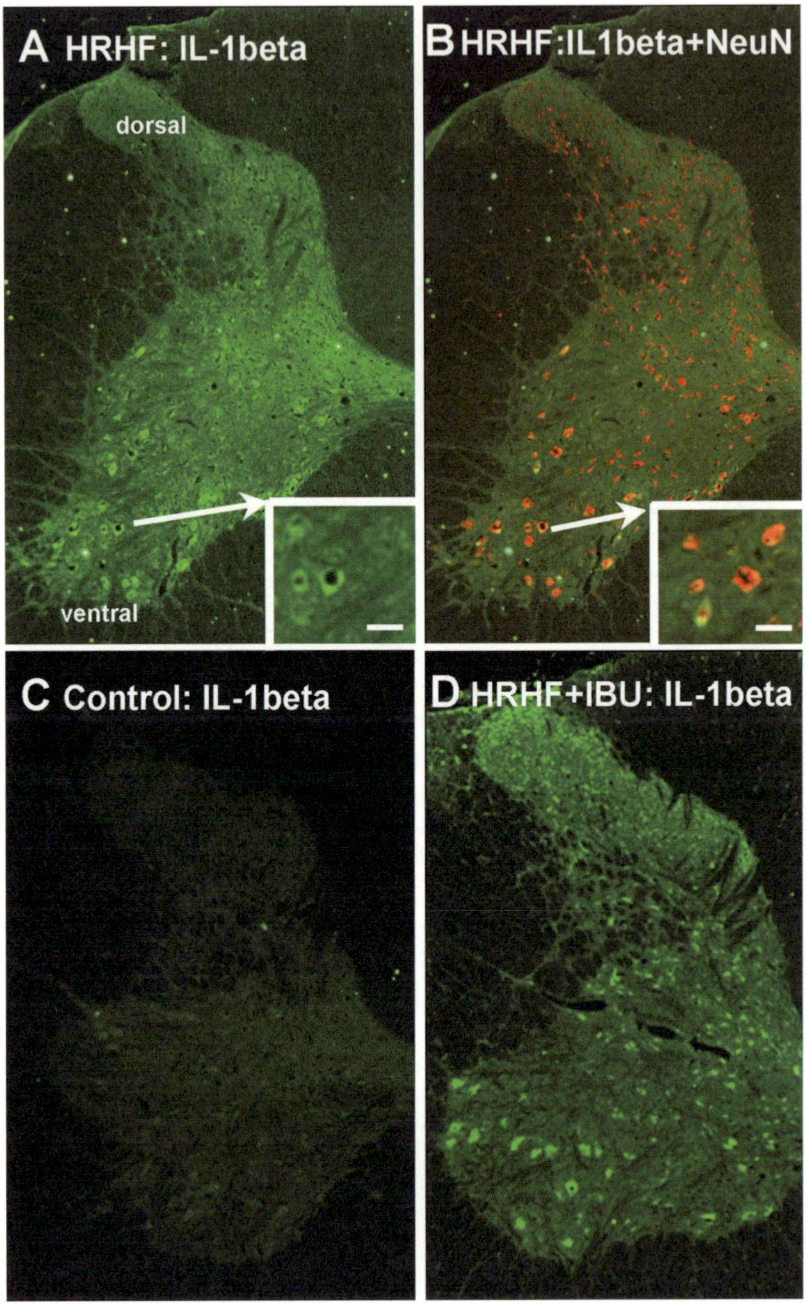

Figure 5. Inflammatory cytokine (IL-1-beta) immunoexpression was increased in neurons of spinal cords of both HRHF and HRHF+IBU rats, compared to NC rats, indicative of central sensitization (n = 4/group, images only shown). (A–D) IL-1-beta immunostained cells that are green in color were visible in spinal cord sections collected from the cervical region (since that region provides input to the median nerve innervating the hand and wrist). These cells were present in the intermediate and ventral horn regions of HRHF rats (A) but none were present in a control rats (C). The red color in panel is NeuN, a neuronal cell body marker. However, IL-1-beta immunostained cells were still visible in spinal cord sections of HRHF+IBU rats (D). Scale bar = 50 μm. Used by permission from Kietrys et al. [41].

http://dx.doi.org/10.5772/66480

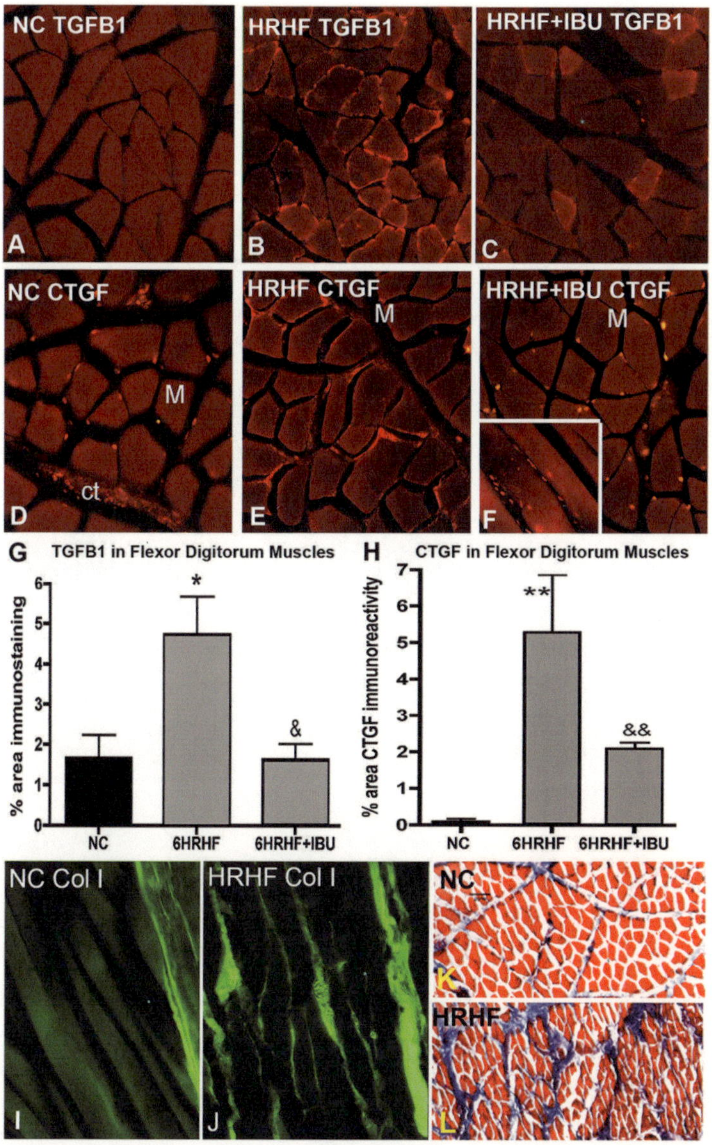

Figure 6. Fibrogenic protein levels (TGFB1, CTGF, and collagen type 1 (Col I)) were increased in forearm muscles of 6-week HRHF rats, increases that were reduced after a 2-week treatment with ibuprofen provided in task weeks 5 and 6. Cross sections of flexor digitorum muscle are shown. (A–C, G) TGFB1 staining was absent in muscles of normal control (NC) rats shown in panel A, high in muscles of untreated 6-week HRHF rats (visible as red staining at the edges of the myofibers in panel B), and reduced in muscles of 6-week HRHF rats treated with ibuprofen (panel C). (D–F, H) A small number of CTGF-immunostained cells (red in color) were present around myofibers in NC rats as shown in panel D, increased in muscles of untreated 6-week HRHF rats as shown in panel E, but reduced back to control levels in muscles of 6-week HRHF rats treated with ibuprofen as shown in panel F. (G&F) Quantification of percentage area of muscle with TGFB1 and CTGF staining. $^{*}p < 0.05$ and $^{**}p < 0.01$, compared to NC rats; $^{\&}p < 0.05$ and $^{\&}p < 0.01$, compared to untreated 6-week HRHF. Scale bars = 50 µm. (I, J) Collagen type 1 (Col I) immunostaining, green in color, is increased considerably between myofibers of 6-week HRHF rats compared to NC rats. These sections were cut longitudinally. (K,L) Another stain (a Masson's trichrome stain, which shows collagen as blue) also shows that collagen deposition is increased between myofibers of 6-week HRHF rats compared to NC rats. Used by permission from Abdelmagid et al. [60].

Eight weeks of ibuprofen administration reduced all of these changes, despite continued task performance (**Figure 7**). This latter finding indicates that the joint degenerative changes observed were a consequence of the inflammatory response induced by this high-repetition high-force task that was 12 weeks in duration. Each of these changes were attenuated by ibuprofen treatment, suggesting that such treatment is chondroprotective, at least during the early phases of cumulative loading-induced inflammation and degeneration in hand and wrist joints.

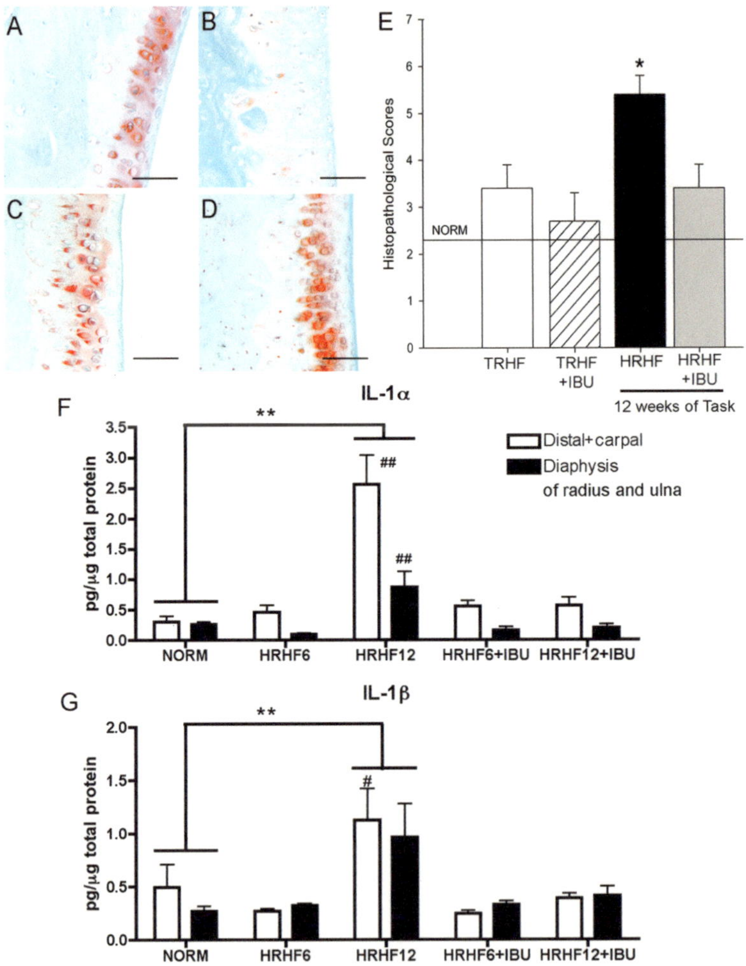

Figure 7. HRHF-induced degeneration of radiocarpal joint cartilage was attenuated by ibuprofen treatment. (A–D) Distal radii articular cartilage stained with safranin O and fast green from (A) untreated TRHF rat, (B) TRHF+IBU rat (trained controls receiving ibuprofen treatment prophylactically), (D) HRHF rats that performed the task for 12 weeks show dramatically reduced proteoglycan staining in the articular cartilage (red-pink safranin O staining), and (E) HRHF+IBU rats that performed the task for 12 weeks while receiving ibuprofen treatment (45 mg/kg body wt, daily, oral) in the last 8 weeks. (E) Histopathological Mankin scores for distal radius articular cartilage of the reach limb in TRHF, TR+IBU, HRHF, and HRHF+IBU rats. (F&G) Cytokine concentrations in wrist joint (distal radius, ulna, and carpal bones) and in diaphysis of the radius and ulna bones, tested using ELISA. Levels of (F) IL-1-alpha and (G) IL-1-beta are shown for each group. $^{*}p < 0.05$ and $^{**}p < 0.01$, compared to NC rats (terms NORM); $^{\#}p < 0.05$ and $^{\#\#}p < 0.01$, compared to untreated 6-week HRHF. Modified with permission from Driban et al. [70], and used by permission.

http://dx.doi.org/10.5772/66480

5.5. Ibuprofen effectively ameliorated osteopenia by reducing task-induced cytokines and osteoclast activity in bones

Cyclical loading and high-force loads are known to affect bone quality [71–74]. However, only a few studies have examined changes occurring in upper extremity bones as a consequence of prolonged performance of occupational tasks. Bone scan studies of patients with upper extremity MSDs show increased blood flow and pooling (suggestive of inflammation) in affected bones, although the sensitivity and accuracy of the results were variable across studies [75–77]. We found that the performance of a HRHF task for 12 weeks reduced trabecular bone (**Figure 8A, B**) and cortical bone thinning in the radius and ulna in untreated HRHF rats (**Figure 8B**) [39, 40]. Bone levels of IL-1-beta, an inflammatory cytokine known to stimulate osteoclastogenesis and activity [78, 79], increased in involved distal forelimb bones (**Figure 8C**). This increase was matched by increased osteoclasts (**Figure 8C**) and increases in two serum biomarkers of bone degradation (Trap5b, band 5 tartrate-resistant acid phosphatase, and a biomarker of osteoclast activity and bone resorption, and CTX1, the C-terminal telopeptide of collagen type I cleaved by osteoclasts during bone resorption). Thus, a 12-week task at high-repetition high-force levels leads to a net loss of trabecular bone volume in the radius and ulna.

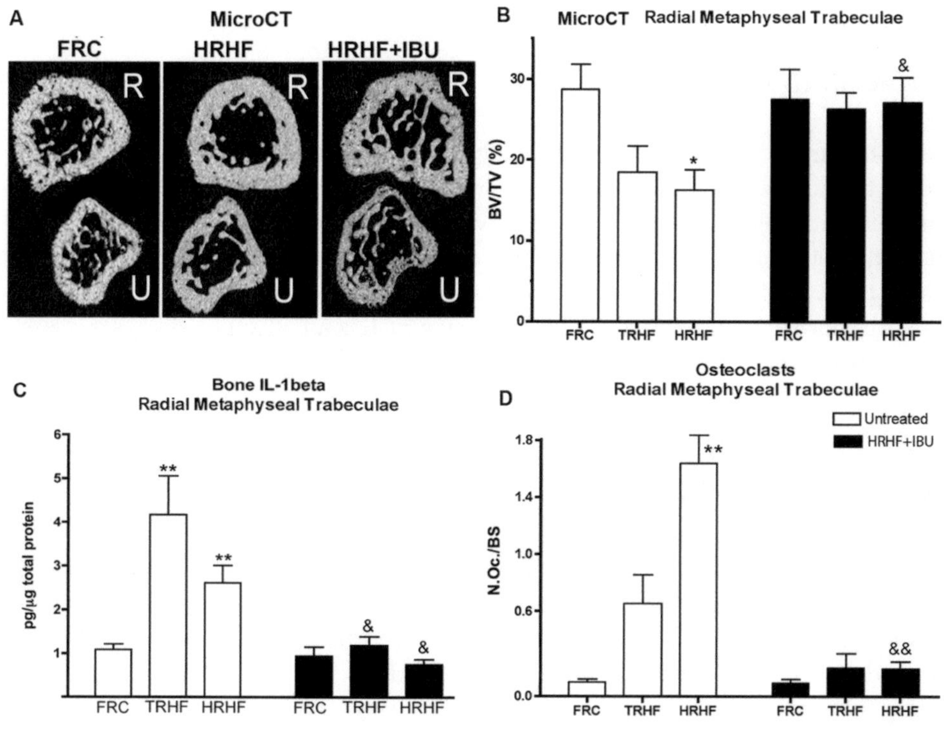

Figure 8. Microcomputed tomography (MicroCT), bone cytokines, and osteoclast numbers in distal radial trabecular region. (A) Representative transaxial microCT slices of the metaphysis of the radius and ulna (at 166 slices, 1.5 mm from the distal edge of their respective growth plate) from an FRC, 12-week HRHF, and 12-week HRHF+IBU rat. (B) MicroCT analysis of trabeculae of distal radius showing reduced trabecular bone volume (BV/TV) in HRHF rats that was rescued by ibuprofen treatment in task weeks 5–12. (C) IL-1-beta in forelimb bones (radius and ulna), tested using ELISA. $^{*}p < 0.05$ and $^{**}p < 0.01$, compared to FRC rats; $^{\&}p < 0.05$, compared to untreated HRHF rats. (D) Density of osteoclasts (N.Oc.), normalized to bone surface (BS), of distal radial metaphyseal trabeculae. Used by permission from Jain et al. [40].

Fortunately, systemic anti-inflammatory treatment with ibuprofen prevented these bone catabolic changes (**Figure 8**) [40, 70]. Eight weeks of continual ibuprofen treatment reduced bone inflammatory cytokine levels, and osteoclast numbers and activity, despite continued task performance. These results suggest that bone catabolism in the untreated HRHF rats was the result of increased inflammatory cytokines and their activating effects on osteoclasts. In summary, forearm bone osteopenia can be one consequence of prolonged high-intensity hand and wrist tasks. This increase in osteopenia and perhaps even fracture risk of workers performing this type of task is under-investigated in human and should be the focus of future studies.

A loss of bone mineral density has been reported in metacarpal bones and distal radius and ulna of patients with long-term carpal tunnel syndrome [80]. Surgical release treatment for carpal tunnel syndrome rescues this decline in distal forearm bone mineral density [81]. Those authors hypothesized that nerve-compression-induced muscle weakness led to bone loss as a consequence of reduced muscular loading on the bone [80], since the muscles involved in performing hand-grip actions produce forces on forearm bones [82, 83]. In our model, ibuprofen may be sparing bone volume by reducing osteoclastogenesis and activity as well as by reducing fibrotic nerve compression, thus sparing muscle activity and muscle-pulling forces on bones (refer to **Figure 3C** and **4A** again).

6. Caveats of ibuprofen use

Ibuprofen treatment is inexpensive and readily available over the counter. Yet, its use should be limited to short-term treatments (we have tested only up to 8 weeks). Ibuprofen medication may inhibit skeletal muscle hypertrophy and adaptation [42, 84–86], although a more recent study shows no effect of ibuprofen on muscle hypertrophy [43]. Long-term use of ibuprofen-related NSIADs could increase gastrointestinal bleeding, renal toxicity, risk of myocardial infarction, and hypertension [87, 88]. NSAIDs are also not always successful for long-term treatment of pain and dysfunction [16], similar to our results with reflexive grip strength.

7. New treatment directions

It is unlikely that a single drug will be effective in treating all WMSDs since their development is multi-factorial. Multipronged treatment should be developed that are individualized to the subject for complete reversal of WMSD-induced tissue inflammation/fibrosis/degeneration and recovery of function. **Figure 9** shows various points of interventional treatment, indicating that early treatment is needed to alter acute inflammatory responses, while chronic inflammatory responses are accompanied by several signs and symptoms of chronic pain and should be treated with secondary anti-inflammatory drugs such as ibuprofen or anti-tumor necrosis factor alpha drugs [89]. The latter drugs have yet to be tested in subjects with WMSDs, but have been tested in our animal model and show fair to strong efficacy [39, 61]. In subjects with chronic or persistent pain, negative neuroplasticity in the CNS, termed central sensitization, may have occurred. Treatment options of such central sensitization should be explored carefully in future studies to reduce chronic pain. At the right side of this figure, we show the onset of fibrosis, which may compress and damage axons (such as in carpal tunnel syndrome), and tether tissues. We are currently exploring options of blocking fibrogenic-signaling pathways in our rat model.

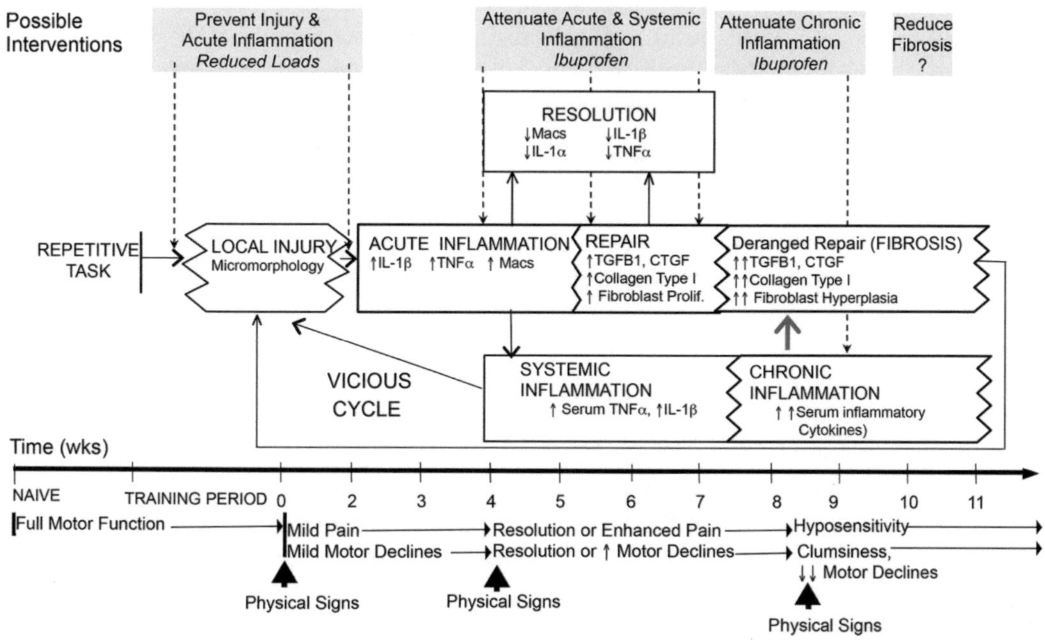

Figure 9. Summary of results and possible points of intervention indicated. Repetitive tasks can cause local injury and acute inflammation that could be prevented with reduced loads. Acute and chronic inflammation can be treated by prophylactic and secondary ibuprofen treatment at anti-inflammatory doses. Means to treat fibrosis are still under investigation as are most effective ways to rescue persistent sensorimotor declines. Abbreviations: CTGF = connective tissue growth factor; IL-1 = interleukin 1; Macs = macrophages; TNF = tumor necrosis factor; TGFB1 = transforming growth factor beta 1. Modified and used by permission from Barr and Barbe [89].

One new non-pharmaceutical direction may be modeled manual therapy. A recent review examined the effectiveness of exercise versus several types of mobilization methods for the treatment of carpal tunnel syndrome and concluded that there was only poor support [90]. However, two recent pilot studies examined massage therapy methods specifically and observed reduced symptoms of discomfort and increased strength post treatment in patients being treated for carpal tunnel syndrome [91, 92]. Another type of massage termed "sports massage' has been used to treat post-exertional muscle soreness, which is also known as delayed onset muscle soreness (DOMS). While the clinical utility of sports massage for DOMS is supported overall, a comprehensive review of the literature by Moraska in 2005 shows its effectiveness in some studies and a lack thereof in others [93]. Perhaps, this is because sports massage therapy treatment is typically short term. With regard to the use of massage therapies for individuals with repetitive motion disorders, clinicians should be aware that these disorders are not acute in nature. Instead, repetitive motion disorders are the consequence of underlying tissue changes that take weeks to years. It is unlikely that a single, short-term treatment will be effective.

Because we could not identify any studies using manual therapies for WMSDs (other than carpal tunnel syndrome), we recently performed a study examining the effectiveness of modeled manual therapy (MMT) as a treatment for symptoms of discomfort, reduced grip strength, and increased tissue fibrosis occurring in forearms of rats performing a HRHF task for 12 weeks [33]. We began the MMT immediately post training to the high-force level, a time point when the rats began to display signs and symptoms consistent with WMSDs. Results were compared to untreated HRHF rats and to age-matched control rats. The MMT protocol included a mixture of manual therapy submodalities: gentle mobilization, skin rolling, and myofascial release (deep massage) of the forearm flexor compartment; joint mobilization of the wrist (gentle rotation and traction of the wrist); and stretching of the entire upper limb from the shoulder to the fingers. The therapy was provided 5 days per week for 12 weeks, while the animals performed the HRHF task for a food reward (as above, for 3 days/week, 2 h/day, in 30-min sessions). Compared to untreated HRHF rats, the HRHF rats receiving the MMT (called HRHF-MMT rats) showed significantly fewer behaviors suggestive of discomfort and had increased numbers of successful reaches. Grip strength had decreased significantly post training to the high-force levels, compared to the rats' naïve levels. However, the MMT protocol improved grip strength within 2 weeks of treatment, an improvement that continued through week 12 despite continued performance of the HRHF task by the HRHF-MMT rats. An examination of tissues post euthanasia showed decreased nerve and connective tissue fibrosis, and decreased collagen and TGF-B1 in the 12-week HRHF rats, compared to the untreated HRHF rats. These observations support further investigation of manual therapy as a preventative for repetitive motion disorders.

Acknowledgements

Research reported in this publication was supported by the National Institutes of Health, the National Institute of Arthritis and Musculoskeletal and Skin Diseases, under Award Numbers AR056019 and AT009350 to MFB.

Author details

Mary F. Barbe[1*] and Ann E. Barr-Gillespsie[2]

*Address all correspondence to: mary.barbe@temple.edu

1 Department of Anatomy and Cell Biology, Lewis Katz School of Medicine at Temple University, Philadelphia, PA, USA

2 College of Health Professions, Pacific University, Hillsboro, OR, USA

References

[1] Woolf AD, Pfleger B. Burden of major musculoskeletal conditions. Bull World Health Organ. 2003;81(9):646–56. Epub 2004/01/09. PubMed PMID: 14710506; PubMed Central PMCID: PMC2572542.

[2] Bureau of Labor Statistics. Nonfatal occupational injuries and illnesses requiring days away from work, 2011. http://www.bls.gov/news.release/osh2.nr0.htm News Release USDL-12-2204, November 8, 2012. 2012.

[3] HSE. The Health and Safety Executive Statistics 2010/11.http://www.hse.gov.uk/statistics/overall/hssh1011.pdf. 2011.

[4] Horton R. GBD 2010: understanding disease, injury, and risk. Lancet. 2012;380(9859):2053–4. Epub 2012/12/19. doi: 10.1016/S0140-6736(12)62133-3. PubMed PMID: 23245595.

[5] Punnett L, Wegman DH. Work-related musculoskeletal disorders: the epidemiologic evidence and the debate. J Electromyogr Kinesiol. 2004;14(1):13–23. Epub 2004/02/05. doi: 10.1016/j.jelekin.2003.09.015. PubMed PMID: 14759746.

[6] Hauret KG, Jones BH, Bullock SH, Canham-Chervak M, Canada S. Musculoskeletal injuries description of an under-recognized injury problem among military personnel. Am J Prev Med. 2010;38(1 Suppl):S61–70. Epub 2010/02/13. doi: 10.1016/j.amepre.2009.10.021. PubMed PMID: 20117601.

[7] Barbe MF, Gallagher S, Massicotte VS, Tytell M, Popoff SN, Barr-Gillespie AE. The interaction of force and repetition on musculoskeletal and neural tissue responses and sensorimotor behavior in a rat model of work-related musculoskeletal disorders. BMC Musculoskelet Disord. 2013;14(1):303. Epub 2013/10/26. doi: 10.1186/1471-2474-14-303. PubMed PMID: 24156755.

[8] Gallagher S, Heberger JR. Examining the interaction of force and repetition on musculoskeletal disorder risk: a systematic literature review. Hum Factors. 2013;55(1):108–24. Epub 2013/03/23. PubMed PMID: 23516797.

[9] Bureau of Labor Statistics. Prevention of work-related musculoskeletal disorders. http://www.bls.gov/news.release/pdf/osh2.pdf. 2014.

[10] Stephens AS, Stephens SR, Morrison NA. Internal control genes for quantitative RT-PCR expression analysis in mouse osteoblasts, osteoclasts and macrophages. BMC Research Notes. 2011;4:410. Epub 2011/10/15. doi: 10.1186/1756-0500-4-410. PubMed PMID: 21996334; PubMed Central PMCID: PMC3204251.

[11] World Health Organization. Workers' health: global plan of action 2007. Available from: http://www.who.int/occupational_health/publications/global_plan/en/.

[12] Clarke J. Organizing the approach to musculoskeletal misuse syndrome. Nurse Pract. 2001;26(7 Pt 1):11–5, 9–25. PubMed PMID: 11494933.

[13] Lashuay N, Burgel B, Harrison R, Israel L, Chan J, Cusac C, et al. We spend our days working in pain: a report on workplace injuries in the garment industry. Final Report for the California Wellness Foundation, Oakland: Asian Immigrant Women Advocates Oakland, CA: Asian Immigrant Women Advocates; 2002 [cited 2016 October 1].

[14] Anthony MJ, Martin EG, Avery AM, Williams JM. Self care and health-seeking behavior of migrant farmworkers. J Immigr Minor Health. 2010;12(5):634–9. Epub 2009/04/25. doi: 10.1007/s10903-009-9252-9. PubMed PMID: 19390972.

[15] Beebe FA, Barkin RL, Barkin S. A clinical and pharmacologic review of skeletal muscle relaxants for musculoskeletal conditions. Am J Ther. 2005;12(2):151–71. Epub 2005/03/16. PubMed PMID: 15767833.

[16] Curatolo M, Bogduk N. Pharmacologic pain treatment of musculoskeletal disorders: current perspectives and future prospects. Clin J Pain. 2001;17(1):25–32. Epub 2001/04/06. PubMed PMID: 11289086.

[17] Krause N, Scherzer T, Rugulies R. Physical workload, work intensification, and prevalence of pain in low wage workers: results from a participatory research project with hotel room cleaners in Las Vegas. Am J Ind Med. 2005;48(5):326–37. Epub 2005/09/30. doi: 10.1002/ajim.20221. PubMed PMID: 16193494.

[18] Boeuf-Cazou O, Lapeyre-Mestre M, Niezborala M, Montastruc JL. Evolution of drug consumption in a sample of French workers since 1986: the 'Drugs and Work' study. Pharmacoepidemiol Drug Saf. 2009;18(4):335–43. doi: 10.1002/pds.1713. PubMed PMID: 19180587.

[19] Graham B. Nonsurgical treatment of carpal tunnel syndrome. J Hand Surg Am. 2009;34(3):531–4. Epub 2009/03/05. doi: 10.1016/j.jhsa.2009.01.010. PubMed PMID: 19258153.

[20] Jarvik JG, Comstock BA, Kliot M, Turner JA, Chan L, Heagerty PJ, et al. Surgery versus non-surgical therapy for carpal tunnel syndrome: a randomised parallel-group trial. Lancet. 2009;374(9695):1074–81. Epub 2009/09/29. doi: 10.1016/S0140-6736(09)61517-8. PubMed PMID: 19782873.

[21] Martin BI, Levenson LM, Hollingworth W, Kliot M, Heagerty PJ, Turner JA, et al. Randomized clinical trial of surgery versus conservative therapy for carpal tunnel syndrome [ISRCTN84286481]. BMC Musculoskelet Disord. 2005;6:2. Epub 2005/01/20. doi: 10.1186/1471-2474-6-2. PubMed PMID: 15656907; PubMed Central PMCID: PMC546190.

[22] O'Connor D, Marshall S, Massy-Westropp N. Non-surgical treatment (other than steroid injection) for carpal tunnel syndrome. Cochrane Database Syst Rev. 2003;(1):CD003219. Epub 2003/01/22. doi: 10.1002/14651858.CD003219. PubMed PMID: 12535461.

[23] Shi Q, MacDermid JC. Is surgical intervention more effective than non-surgical treatment for carpal tunnel syndrome? A systematic review. J Orthop Surg Res. 2011;6:17. Epub 2011/04/12. doi: 10.1186/1749-799X-6-17. PubMed PMID: 21477381; PubMed Central PMCID: PMC3080334.

[24] Verdugo RJ, Salinas RA, Castillo JL, Cea JG. Surgical versus non-surgical treatment for carpal tunnel syndrome. Cochrane Database Syst Rev. 2008;(4):CD001552. Epub 2008/10/10. doi: 10.1002/14651858.CD001552.pub2. PubMed PMID: 18843618.

[25] Rainsford KD. Ibuprofen: A Critical Bibliographic Review. Philadelphia, PA: Taylor and Francis Limited; 1999.

[26] Miller SB. Prostaglandins in health and disease: an overview. Semin Arthritis Rheum. 2006;36(1):37–49. doi: 10.1016/j.semarthrit.2006.03.005. PubMed PMID: 16887467.

[27] Byl NN. Learning-based animal models: task-specific focal hand dystonia. ILAR Journal/ National Research Council, Institute of Laboratory Animal Resources. 2007;48(4):411–31. Epub 2007/08/23. PubMed PMID: 17712226.

[28] Topp KS, Byl NN. Movement dysfunction following repetitive hand opening and closing: anatomical analysis in Owl monkeys. Mov Disord. 1999;14(2):295–306. Epub 1999/03/26. PubMed PMID: 10091624.

[29] Sommerich CM, Lavender SA, Buford JA, J JB, Korkmaz SV, Pease WS. Towards development of a nonhuman primate model of carpal tunnel syndrome: performance of a voluntary, repetitive pinching task induces median mononeuropathy in Macaca fascicularis. J Orthop Res. 2007;25(6):713–24. Epub 2007/02/24. doi: 10.1002/jor.20363. PubMed PMID: 17318891.

[30] Barbe MF, Elliott MB, Abdelmagid SM, Amin M, Popoff SN, Safadi FF, et al. Serum and tissue cytokines and chemokines increase with repetitive upper extremity tasks. J Orthop Res. 2008;26(10):1320–6. Epub 2008/05/09. doi: 10.1002/jor.20674. PubMed PMID: 18464247.

[31] Fisher PW, Zhao Y, Rico MC, Massicotte VS, Wade CK, Litvin J, et al. Increased CCN2, substance P and tissue fibrosis are associated with sensorimotor declines in a rat model of repetitive overuse injury. J Cell Commun Signal. 2015:1–18. doi: 10.1007/ s12079-015-0263-0.

[32] Gao HG, Fisher PW, Lambi AG, Wade CK, Barr-Gillespie AE, Popoff SN, et al. Increased serum and musculotendinous fibrogenic proteins following persistent low-grade inflammation in a rat model of long-term upper extremity overuse. PLoS One. 2013;8(8):e71875. Epub 2013/09/10. doi: 10.1371/journal.pone.0071875. PubMed PMID: 24015193; PubMed Central PMCID: PMC3756034.

[33] Bove GM, Harris MY, Zhao H, Barbe MF. Manual therapy as an effective treatment for fibrosis in a rat model of upper extremity overuse injury. J Neurol Sci. 2016;361:168–80. doi: 10.1016/j.jns.2015.12.029. PubMed PMID: 26810536; PubMed Central PMCID: PMCPMC4729290.

[34] Barr AE, Barbe MF. Pathophysiological tissue changes associated with repetitive movement: a review of the evidence. Phys Ther. 2002;82(2):173–87. Epub 2002/02/22. PubMed PMID: 11856068; PubMed Central PMCID: PMC1550512.

[35] Barbe MF, Barr AE, Gorzelany I, Amin M, Gaughan JP, Safadi FF. Chronic repetitive reaching and grasping results in decreased motor performance and widespread tissue responses in a rat model of MSD. J Orthop Res. 2003;21(1):167–76. Epub 2003/01/01. doi: 10.1016/S0736-0266(02)00086-4. PubMed PMID: 12507595; PubMed Central PMCID: PMC1560095.

[36] Clark BD, Al Shatti TA, Barr AE, Amin M, Barbe MF. Performance of a high-repetition, high-force task induces carpal tunnel syndrome in rats. J Orthop Sports PhysTher. 2004;34(5):244–53.

[37] Barbe MF, Gallagher S, Popoff SN. Serum biomarkers as predictors of stage of work-related musculoskeletal disorders. J Am Acad Orthop Surg. 2013;21(10):644–6. Epub 2013/10/03. doi: 10.5435/JAAOS-21-10-644. PubMed PMID: 24084439.

[38] Adams SS, Bough RG, Cliffe EE, Dickinson W, Lessel B, McCullough KF, et al. Some aspects of the pharmacology, metabolism, and toxicology of ibuprofen. I. Pharmacology and metabolism. Rheumatol Phys Med. 1970;10:Suppl 10:9–26. Epub 1970/01/01. PubMed PMID: 5524289.

[39] Rani S, Barbe MF, Barr AE, Litivn J. Role of TNF alpha and PLF in bone remodeling in a rat model of repetitive reaching and grasping. J Cell Physiol. 2010;225(1):152–67. Epub 2010/05/12. doi: 10.1002/jcp.22208. PubMed PMID: 20458732.

[40] Jain NX, Barr-Gillespie AE, Clark BD, Kietrys DM, Wade CK, Litvin J, et al. Bone loss from high repetitive high force loading is prevented by ibuprofen treatment. J Musculoskelet Neuronal Interact. 2014;14(1):78–94. PubMed PMID: 24583543; PubMed Central PMCID: PMCPMC4067254.

[41] Kietrys DM, Barr AE, Barbe MF. Exposure to repetitive tasks induces motor changes related to skill acquisition and inflammation in rats. J Mot Behav. 2011;43(6):465–76. Epub 2011/11/18. doi: 10.1080/00222895.2011.627897. PubMed PMID: 22087754.

[42] Trappe TA, Carroll CC, Dickinson JM, LeMoine JK, Haus JM, Sullivan BE, et al. Influence of acetaminophen and ibuprofen on skeletal muscle adaptations to resistance exercise in older adults. Am J Physiol Regul Integr Comp Physiol. 2011;300(3):R655–62. Epub 2010/12/17. doi: ajpregu.00611.2010 [pii] 10.1152/ajpregu.00611.2010. PubMed PMID: 21160058; PubMed Central PMCID: PMC3064281.

[43] Krentz JR, Quest B, Farthing JP, Quest DW, Chilibeck PD. The effects of ibuprofen on muscle hypertrophy, strength, and soreness during resistance training. Appl Physiol Nutr Metab. 2008;33(3):470–5. Epub 2008/05/08. doi: h08-019 [pii] 10.1139/H08-019. PubMed PMID: 18461099.

[44] Schafers M, Sorkin LS, Sommer C. Intramuscular injection of tumor necrosis factor-alpha induces muscle hyperalgesia in rats. Pain. 2003;104(3):579–88. Epub 2003/08/21. doi: S0304395903001155 [pii]. PubMed PMID: 12927630.

[45] Stauber WT, Clarkson PM, Fritz VK, Evans WJ. Extracellular matrix disruption and pain after eccentric muscle action. J Appl Physiol. 1990;69(3):868–74. Epub 1990/09/01. PubMed PMID: 2123179.

[46] Stauber WT, Knack KK, Miller GR, Grimmett JG. Fibrosis and intercellular collagen connections from four weeks of muscle strains. Muscle Nerve. 1996;19(4):423–30. Epub 1996/04/01. doi: 10.1002/mus.880190402. PubMed PMID: 8622719.

[47] Stauber WT, Smith CA, Miller GR, Stauber FD. Recovery from 6 weeks of repeated strain injury to rat soleus muscles. Muscle Nerve. 2000;23(12):1819–25. Epub 2000/12/05. PubMed PMID: 11102904.

[48] Willems ME, Miller GR, Stauber FD, Stauber WT. Effects of repeated lengthening contractions on skeletal muscle adaptations in female rats. J Physiol Sci. 2010;60(2):143–50. Epub 2010/01/07. doi: 10.1007/s12576-009-0078-y. PubMed PMID: 20052570.

[49] Smith CA, Stauber F, Waters C, Alway SE, Stauber WT. Transforming growth factor-beta following skeletal muscle strain injury in rats. J Appl Physiol. 2007;102(2):755–61. Epub 2006/10/28. doi: 01503.2005 [pii] 10.1152/japplphysiol.01503.2005. PubMed PMID: 17068209.

[50] Nakama LH, King KB, Abrahamsson S, Rempel DM. VEGF, VEGFR-1, and CTGF cell densities in tendon are increased with cyclical loading: An in vivo tendinopathy model. J Orthop Res. 2006;24(3):393–400. Epub 2006/02/16. doi: 10.1002/jor.20053. PubMed PMID: 16479573.

[51] Shen W, Li Y, Zhu J, Schwendener R, Huard J. Interaction between macrophages, TGF-beta1, and the COX-2 pathway during the inflammatory phase of skeletal muscle healing after injury. J Cell Physiol. 2008;214(2):405–12. Epub 2007/07/28. doi: 10.1002/jcp.21212. PubMed PMID: 17657727.

[52] Garrett Q, Khaw PT, Blalock TD, Schultz GS, Grotendorst GR, Daniels JT. Involvement of CTGF in TGF-beta1-stimulation of myofibroblast differentiation and collagen matrix contraction in the presence of mechanical stress. Invest Ophthalmol Vis Sci. 2004;45(4):1109–16. Epub 2004/03/24. PubMed PMID: 15037576.

[53] Heinemeier KM, Olesen JL, Haddad F, Langberg H, Kjaer M, Baldwin KM, et al. Expression of collagen and related growth factors in rat tendon and skeletal muscle in response to specific contraction types. J Physiol. 2007;582(Pt 3):1303–16. Epub 2007/06/02. doi: jphysiol.2007.127639 [pii] 10.1113/jphysiol.2007.127639. PubMed PMID: 17540706; PubMed Central PMCID: PMC2075262.

[54] Daniels JT, Schultz GS, Blalock TD, Garrett Q, Grotendorst GR, Dean NM, et al. Mediation of transforming growth factor-beta(1)-stimulated matrix contraction by fibroblasts: a role for connective tissue growth factor in contractile scarring. Am J Pathol. 2003;163(5):2043–52. PubMed PMID: 14578203; PubMed Central PMCID: PMC1892432.

[55] Guo F, Carter DE, Leask A. Mechanical tension increases CCN2/CTGF expression and proliferation in gingival fibroblasts via a TGFbeta-dependent mechanism. PLoS One. 2011;6(5):e19756. Epub 2011/05/26. doi: 10.1371/journal.pone.0019756 PONE-D-11-01727 [pii]. PubMed PMID: 21611193; PubMed Central PMCID: PMC3096639.

[56] Yu F, Chou CW, Chen CC. TNF-alpha suppressed TGF-beta-induced CTGF expression by switching the binding preference of p300 from Smad4 to p65. Cell Signal. 2009;21(6):867–72. Epub 2009/04/23. PubMed PMID: 19385047.

[57] Beddy D, Mulsow J, Watson RW, Fitzpatrick JM, O'Connell PR. Expression and regulation of connective tissue growth factor by transforming growth factor beta and tumour necrosis factor alpha in fibroblasts isolated from strictures in patients with Crohn's disease. Br J Surg. 2006;93(10):1290–6. Epub 2006/07/14. doi: 10.1002/bjs.5431. PubMed PMID: 16838391.

[58] Rani S, Barbe MF, Barr AE, Litvin J. Periostin-like-factor and Periostin in an animal model of work-related musculoskeletal disorder. Bone. 2009;44(3):502–12. Epub 2008/12/20. doi: 10.1016/j.bone.2008.11.012. PubMed PMID: 19095091.

[59] Clark BD, Barr AE, Safadi FF, Beitman L, Al-Shatti T, Amin M, et al. Median nerve trauma in a rat model of work-related musculoskeletal disorder. J Neurotrauma. 2003;20(7):681–95. Epub 2003/08/12. doi: 10.1089/089771503322144590. PubMed PMID: 12908929; PubMed Central PMCID: PMC1550513.

[60] Fedorczyk JM, Barr AE, Rani S, Gao HG, Amin M, Amin S, et al. Exposure-dependent increases in IL-1beta, substance P, CTGF, and tendinosis in flexor digitorum tendons with upper extremity repetitive strain injury. J Orthop Res. 2010;28(3):298–307. Epub 2009/09/11. doi: 10.1002/jor.20984. PubMed PMID: 19743505; PubMed Central PMCID: PMC2807907.

[61] Abdelmagid SM, Barr AE, Rico M, Amin M, Litvin J, Popoff SN, et al. Performance of repetitive tasks induces decreased grip strength and increased fibrogenic proteins in skeletal muscle: role of force and inflammation. PLoS One. 2012;7(5):e38359. Epub 2012/06/08. doi: 10.1371/journal.pone.0038359. PubMed PMID: 22675458; PubMed Central PMCID: PMC3364991.

[62] Brandt KD, Dieppe P, Radin EL. Etiopathogenesis of osteoarthritis. Rheum Dis Clin North Am. 2008;34(3):531–59. Epub 2008/08/09. doi: S0889-857X(08)00045-8 [pii] 10.1016/j.rdc.2008.05.011. PubMed PMID: 18687271.

[63] Calfee RP, Leventhal EL, Wilkerson J, Moore DC, Akelman E, Crisco JJ. Simulated radioscapholunate fusion alters carpal kinematics while preserving dart-thrower's motion. J Hand Surg Am. 2008;33(4):503–10. Epub 2008/04/15. doi: S0363-5023(07)01104-5 [pii] 10.1016/j.jhsa.2007.12.013. PubMed PMID: 18406953.

[64] Martin JA, Buckwalter JA. Post-traumatic osteoarthritis: the role of stress induced chondrocyte damage. Biorheology. 2006;43(3-4):517–21. Epub 2006/08/17. PubMed PMID: 16912423.

[65] Bernard B. Musculoskeletal disorders and workplace factors. NIOSH Report 97–141 Cincinatti (OH): National Institute for Occupational Safety and Health 1997.

[66] Solovieva S, Vehmas T, Riihimaki H, Luoma K, Leino-Arjas P. Hand use and patterns of joint involvement in osteoarthritis. A comparison of female dentists and teachers. Rheumatology (Oxford). 2005;44(4):521–8. Epub 2005/02/25. doi: keh534 [pii] 10.1093/rheumatology/keh534. PubMed PMID: 15728421.

[67] Ding H, Solovieva S, Vehmas T, Takala EP, Leino-Arjas P. Hand osteoarthritis and pinch grip strength among middle-aged female dentists and teachers. Scand J Rheumatol. 2010;39(1):84–7. Epub 2010/02/06. doi: 10.3109/03009740903201834. PubMed PMID: 20132076.

[68] Dominick KL, Jordan JM, Renner JB, Kraus VB. Relationship of radiographic and clinical variables to pinch and grip strength among individuals with osteoarthritis. Arthritis Rheum. 2005;52(5):1424–30. Epub 2005/05/10. doi: 10.1002/art.21035. PubMed PMID: 15880347.

[69] Spies-Dorgelo MN, van der Windt DA, van der Horst HE, Prins AP, Stalman WA. Hand and wrist problems in general practice—patient characteristics and factors related to symptom severity. Rheumatology (Oxford). 2007;46(11):1723–8. Epub 2007/10/17. doi: kem253 [pii] 10.1093/rheumatology/kem253. PubMed PMID: 17938132.

[70] Driban JB, Barr AE, Amin M, Sitler MR, Barbe MF. Joint inflammation and early degeneration induced by high-force reaching are attenuated by ibuprofen in an animal model of work-related musculoskeletal disorder. J Biomed Biotechnol. 2011;2011:691412. Epub 2011/03/16. doi: 10.1155/2011/691412. PubMed PMID: 21403884; PubMed Central PMCID: PMC3051200.

[71] Rubin C, Turner AS, Muller R, Mittra E, McLeod K, Lin W, et al. Quantity and quality of trabecular bone in the femur are enhanced by a strongly anabolic, noninvasive mechanical intervention. J Bone Miner Res. 2002;17(2):349–57. Epub 2002/01/29. doi: 10.1359/jbmr.2002.17.2.349. PubMed PMID: 11811566.

[72] Gross TS, Srinivasan S. Building bone mass through exercise: could less be more? Br J Sports Med. 2006;40(1):2–3; discussion 2–3. Epub 2005/12/24. doi: 10.1136/bjsm.2004.016972. PubMed PMID: 16371481; PubMed Central PMCID: PMC1435732.

[73] Srinivasan S, Weimer DA, Agans SC, Bain SD, Gross TS. Low-magnitude mechanical loading becomes osteogenic when rest is inserted between each load cycle. J Bone Miner Res. 2002;17(9):1613–20. Epub 2002/09/05. doi: 10.1359/jbmr.2002.17.9.1613. PubMed PMID: 12211431; PubMed Central PMCID: PMC1435731.

[74] Bourrin S, Genty C, Palle S, Gharib C, Alexandre C. Adverse effects of strenuous exercise: a densitometric and histomorphometric study in the rat. J Appl Physiol. 1994;76(5):1999–2005. Epub 1994/05/01. PubMed PMID: 8063662.

[75] Amorim BJ, Etchebehere EC, Dalla Torre G, Lima Mda C, Santos Ade O, Ramos CD, et al. Low sensitivity of three-phase bone scintigraphy for the diagnosis of repetitive strain injury. Sao Paulo Med J. 2006;124(3):145–9. Epub 2006/11/23. PubMed PMID: 17119691.

[76] Al-Nahhas AM, Jawad AS, McCready VR, Kedar R. Detection of increased blood flow to the affected arm in repetitive strain injury with radionuclide and Doppler ultrasound studies. A case report. Clin Nucl Med. 1995;20(7):615–8. Epub 1995/07/01. PubMed PMID: 7554665.

[77] Al-Nahhas AM, Jawad AS, Norman A, McCready VR. 99Tcm-MDP blood-pool phase in the assessment of repetitive strain injury. Nucl Med Commun. 1997;18(10):927–31. Epub 1997/12/11. PubMed PMID: 9392793.

[78] Kulkarni RN, Bakker AD, Everts V, Klein-Nulend J. Mechanical loading prevents the stimulating effect of IL-1beta on osteocyte-modulated osteoclastogenesis. Biochem Biophys Res Commun. 2012;420(1):11–6. Epub 2012/03/07. doi: 10.1016/j.bbrc.2012.02.099. PubMed PMID: 22390927.

[79] Nanes MS. Tumor necrosis factor-alpha: molecular and cellular mechanisms in skeletal pathology. Gene. 2003;321:1–15. Epub 2003/11/26. PubMed PMID: 14636987.

[80] Erselcan T, Topalkara K, Nacitarhan V, Akyuz A, Dogan D. Carpal tunnel syndrome leads to significant bone loss in metacarpal bones. J Bone Miner Metab. 2001;19(5):317–20. Epub 2001/08/11. doi: 10.1007/s0077410190317. PubMed PMID: 11498735.

[81] Schorn D, Hoskinson J, Dickson RA. Bone density and the carpal tunnel syndrome. Hand. 1978;10(2):184–6. Epub 1978/06/01. PubMed PMID: 711001.

[82] Hasegawa Y, Schneider P, Reiners C. Age, sex, and grip strength determine architectural bone parameters assessed by peripheral quantitative computed tomography (pQCT) at the human radius. J Biomech. 2001;34(4):497–503. Epub 2001/03/27. PubMed PMID: 11266673.

[83] Desrosiers J, Hebert R, Bravo G, Rochette A. Age-related changes in upper extremity performance of elderly people: a longitudinal study. Exp Gerontol. 1999;34(3):393–405. Epub 1999/08/05. PubMed PMID: 10433393.

[84] Machida M, Takemasa T. Ibuprofen administration during endurance training cancels running-distance-dependent adaptations of skeletal muscle in mice. J Physiol Pharmacol. 2010;61(5):559–63. Epub 2010/11/18. PubMed PMID: 21081799.

[85] Soltow QA, Betters JL, Sellman JE, Lira VA, Long JH, Criswell DS. Ibuprofen inhibits skeletal muscle hypertrophy in rats. Med Sci Sports Exerc. 2006;38(5):840–6. Epub 2006/05/05. doi: 10.1249/01.mss.0000218142.98704.66. 00005768-200605000-00006 [pii]. PubMed PMID: 16672835.

[86] Ziltener JL, Leal S, Fournier PE. Non-steroidal anti-inflammatory drugs for athletes: an update. Ann Phys Rehab Med. 2010;53(4):278–82, 82–8. Epub 2010/04/07. doi: S1877-0657(10)00057-6 [pii]. 10.1016/j.rehab.2010.03.001. PubMed PMID: 20363203.

[87] Moore, R.A., S. Derry, and H.J. McQuay, *Cyclo-oxygenase-2 selective inhibitors and non-steroidal anti-inflammatory drugs: balancing gastrointestinal and cardiovascular risk.* BMC Musculoskelet Disord, 2007. **8**: p. 73.

[88] Al-Saeed, A., *Gastrointestinal and Cardiovascular Risk of Nonsteroidal Anti-inflammatory Drugs*. Oman Med J, 2011. **26**(6): p. 385–91.

[89] Barr, A.E. and M.F. Barbe, Inflammation reduces physiological tissue tolerance in the development of work-related musculoskeletal disorders. Journal of electromyography and kinesiology: official journal of the International Society of Electrophysiological Kinesiology, 2004. 14(1): p. 77–85.

[90] Page MJ, O'Connor D, Pitt V, Massy-Westropp N. Exercise and mobilisation interventions for carpal tunnel syndrome. Cochrane Database Syst Rev. 2012;6:Cd009899. Epub 2012/06/15. doi: 10.1002/14651858.cd009899. PubMed PMID: 22696387.

[91] Elliott R, Burkett B. Massage therapy as an effective treatment for carpal tunnel syndrome. J Bodyw Mov Ther. 2013;17(3):332–8. Epub 2013/06/19. doi: 10.1016/j.jbmt.2012.12.003. PubMed PMID: 23768278.

[92] Moraska A, Chandler C, Edmiston-Schaetzel A, Franklin G, Calenda EL, Enebo B. Comparison of a targeted and general massage protocol on strength, function, and symptoms associated with carpal tunnel syndrome: a randomized pilot study. J Altern Complement Med. 2008;14(3):259–67. Epub 2008/03/29. doi: 10.1089/acm.2007.0647. PubMed PMID: 18370581.

[93] Moraska A. Sports massage. A comprehensive review. J Sports Med Phys Fitness. 2005;45(3):370–80. Epub 2005/10/19. PubMed PMID: 16230990.

Chapter 10

Work-Related Musculoskeletal Disorders and the Relationship to Ethnicity

Rosana M. Quintana

Additional information is available at the end of the chapter

http://dx.doi.org/10.5772/67002

Abstract

Work-related musculoskeletal disorders (MSDs) are a constellation of painful disorders, which could cause chronic disability. Multiple risk factors, both occupational and non occupational, may be involved. MSDs have been extensively studied in several countries; however, few studies have been carried out based on the relationship between MSDs and ethnicity.

Objective: To describe the relationship between MSDs and ethnicity in different parts of the world.

Methods: A nonsystematic literature review of studies, with both quantitative and qualitative methodology, was conducted.

Results: The evidence in research, qualitative and quantitative, emphasizes the importance of this problem in vulnerable populations in different parts of the world, prioritizing migration to developed countries and precarity condition of work.

Conclusion: Recommendations for a multidisciplinary approach of MSDs in vulnerable groups were raised.

Keywords: work-related musculoskeletal disorders, ethnicity, vulnerable and indigenous populations

1. Introduction

Work-related musculoskeletal disorders (MSDs) are a constellation of painful disorders of muscles, tendons, joints and nerves, which can affect all body parts, although neck, upper limbs and back are the most common areas. Upper extremity musculoskeletal disorders are also highly prevalent in manual-intensive occupations, and back and lower limb disorders occur disproportionately among truck drivers, warehouse workers, construction trades, among others. In most cases, it could cause chronic disability [1, 2].

The appropriate term is "work-related" disorders, as distinguished from specifically "occupational" disorders where a single factor is both, necessary and sufficient, in order to cause the disease [3].

MSDs have multiple risk factors, both occupational and nonoccupational. In addition to work demands, other aspects of daily life can influence their appearance. Musculoskeletal tissues can also be affected by systemic diseases [1, 2].

Risk varies by age, gender, socioeconomic status and ethnicity. Other suspected risk factors may include obesity, smoking, muscle strength and other aspects of work capacity [1].

MSDs cause a huge socioeconomic burden to patients and their household, society and their country indeed. Yet, their relevance is often minimized, particularly in developing countries with fragmented healthcare system and poor nations.

MSDs have been extensively studied in several countries; however, few studies have been carried out so as to investigate the relationship between MSDs and ethnicity [4–9].

The main objective of this chapter is to describe the relationship between MSDs and ethnicity in different parts of the world.

2. Material and methods

A nonsystematic literature review of studies, with both quantitative and qualitative methodology, was conducted. Methodological phases proposed by Greenhalgh et al. [10] were in the following (**Figure 1**):

1. Planning phase: It defines the research question for the development of the review.
2. Search phase: The search was conducted in the databases: MEDLINE, EMBASE, LILACS, SOCIAL SCIENCE INDEX and PSYCOINFO. The key words were as follows: work-related musculoskeletal disorders, ethnicity, vulnerable and indigenous populations. A hand search was carried out as well. Studies, in English and Spanish, were included in the current search.
3. Mapping phase: The key words of the selected studies were identified.
4. Synthesis phase: Data relevant to the objectives of this study were synthesized using interpretive analysis.
5. Recommendation phase: The paper summarized all the aspects related to its objective, and since then, recommendations were developed.

http://dx.doi.org/10.5772/67002

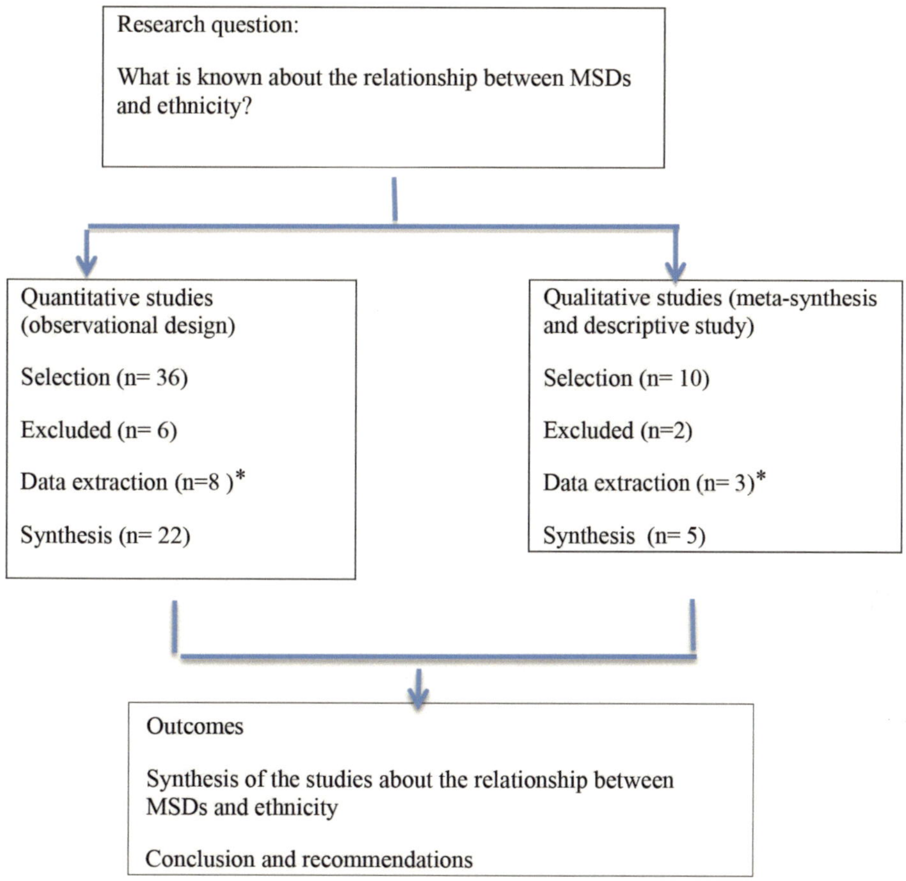

*Articles indirectly relevant to the problem at hand. Not strictly meeting the inclusion criteria.

Figure 1. Methods of review. Greenhalgh et al. [10].

3. Results

Our findings fall into two categories: 1. Summary and analysis of quantitative studies and 2. Synthesis and interpretation of qualitative studies

1. Summary and analysis of quantitative studies (**Table 1**): The literature is mostly of European and American origin. They emphasize that certain population groups, especially immigrants and those belonging to lower socioeconomic status, have more MSDs.

In 1998, Urwin et al. [11] describes that people who live in socially deprived areas in United Kingdom (UK) have more musculoskeletal symptoms. Mergler [12] in Canada prioritizes the combination of qualitative and quantitative methods for this complex problem.

Author, year [ref]	Objectives and method	Country	Group studied	Conclusions
Urwin et al, 1998 [11]	Estimate the frequency of musculoskeletal pain in the adult population. Population survey	UK	Adults	Musculoskeletal pain is common in the community. People who live in socially deprived areas have more musculoskeletal symptoms
Mergler, 1999 [12]	Identify work conditions that affect human health and well-being with a view to reducing or preventing MSDs Combining qualitative and quantitative methods	Canada	Adults	Combining the results allowed us to identify the behaviors and policies that resulted from workplace changes and improved health
Cole et al., 2001 [13]	Describe the prevalence of MSDs in the Canadian working population The Canadian 1994 national population health survey (NPHS)	Canada	Adults household residents	Associations between work stressors and MSDs provide evidence for physical and psychosocial factors both affecting disability in a working population
Vindigni et al., 2004 [14]	Describe the prevalence of MSK disorders in rural indigenous Australians areas A cross-sectional research	Australia	Indigenous members of the community (Kempsey District, New South Wales)	The 57% people have learnt to live with chronic levels of pain affecting multiple anatomical sites
Turner et al., 2004 [15]	Develop statistical models that accurately predict chronic work disability This is a population-based, prospective study	US	Worker interview	The combination of sociodemographic, biomedical, work-related, administrative/legal and psychosocial risk factors for predicting chronic disability in workers
Ahonen et al., 2007 [16]	Summarizes the information on immigrant occupational health available from recent studies, incorporating varied study designs Review	US, Europe, Canada, Asia	Immigrant adults	The immigrations were associated with occupational risk factors, health consequences and the social, economic and cultural influences on worker health
Côté et al., 2008 [17]	Describe the prevalence and incidence of neck pain and disability in workers Review	Canada	Adult workers	Neck pain is endemic in workers throughout the industrialized world

Author, year [ref]	Objectives and method	Country	Group studied	Conclusions
Joshi and Chopra, 2009 [18]	Estimate urban prevalence of MSK disorders and compare to an earlier rural regional study COPCORD methodology	India	Pune adult residents	Similarly, the prevalence of MSK disorders was significantly lower in the urban (current Pune) vs. rural (Bhigwan)
Davatchi et al., 2009 [19]	Compare the prevalence of MSK disorders in Caucasians and Turks in an identical environment COPCORD methodology	Iran	Tehran adult residents	Musculoskeletal complaints were more frequent in Turks than in Caucasians, and the prevalence of rheumatic disorders was rather similar except for knee osteoarthritis
Jørgensen et al., 2011 [7]	Investigate differences self-reported health measures between immigrant and Danish cleaners Cross-sectional study	Denmark	Adults immigrant and Danish	The immigrant cleaners generally had a poorer self-reported health and work ability than the Danish cleaners
Schulz et al., 2013 [5]	Describe the prevalence of upper body musculoskeletal symptoms reported by Latino poultry processing workers and a comparison population of Latino manual workers in Western North Carolina Cross-sectional study	US	Latino poultry workers and other Latino manual workers living in communities	Back symptoms and wrist/hand symptoms were reported by over 35% of workers. Workplace conditions facing poultry processing and indigenous language speaking workers deserve further exploration
Rosenbaum et al., 2013 [6]	Improve understanding of immigrant Latino manual workers' occupational health, focusing on upper body musculoskeletal injury Cross-sectional study	US	Latino poultry workers and other Latino manual workers living in communities	Upper body musculoskeletal and low back pain are common in immigrant Latino workers and may negatively impact long-term health and contribute to occupational health disparities
Xiao et al., 2013 [8]	Characterize the association between agricultural work and chronic musculoskeletal pain. Cross-sectional study	US	Adult workers in California Latino farm	Chronic musculoskeletal pain is prevalent among farm workers and is associated with common work positions

Author, year [ref]	Objectives and method	Country	Group studied	Conclusions
Cartwright et al., 2013 [9]	Determine the incidence of CTS over 1 year in Latino poultry processing workers Community-based sampling.	US	Adult Latinos in poultry and non-poultry manual labor occupations	Latino poultry processing workers have an incidence of CTS that is possibly higher than Latinos in other manual labor positions
Peláez-Ballestas et al., 2015 [20]	Estimate the prevalence of MSK disorders and rheumatic diseases in indigenous Maya-Yucateco COPCORD methodology	Mexico	Adult identified as indigenous resident in Chankom (state of Yucatan)	MSK pain and rheumatic diseases were highly prevalent, with high impact on daily activities
Rodriguez-Amado et al., 2016 [21]	Identify individual and contextual factors associated with the variation of prevalence of OA in the Mexican population Multilevel analysis SGIx was associated with OA	Mexico	Adult identified as indigenous resident in Chankom who had symptomatic OA	These factors were independently associated with the prevalence of OA: female, pain intensity, physical limitation and the use of pain treatments with OA. The association between OA prevalence and regional variations with SGIx reflects inequities in health
Quintana et al., 2016 [22]	Estimate the prevalence of MSK disorders and rheumatic diseases among the indigenous Qom (Toba) in Argentina COPCORD methodology	Argentina	Adult identified as indigenous resident in Rosario (Argentina)	MSK pain and rheumatic diseases were highly prevalent, with high impact on daily activities
Loyola-Sanchez et al., 2016 [23]	Evaluate the impact of arthritis on the physical function of people living in a Maya Yucateco rural community COPCORD methodology	Mexico	Adult identified as indigenous resident in Chankom with arthritis	Arthritis is highly associated with disability. The prevalence of arthritis is associated with social factors, in addition to individual factors
Julián-Santiago et al., 2016 [24]	Estimate the prevalence of MSK disorders and rheumatic diseases in the Chontal and Mixtec indigenous communities in Mexico COPCORD methodology	Mexico	Adult identified as Mixtec and Chontal indigenous resident in rural areas of Oaxaca	The prevalence of MSK disorders was 45.5 %. The most common rheumatic diseases were back pain and osteoarthritis A high percentage of participants had not received medical care

Author, year [ref]	Objectives and method	Country	Group studied	Conclusions
Del Río Nájera et al., 2016 [25]	Determine the prevalence of MSK pain and rheumatic diseases in the Raramuri indigenous (Chihuahua) in Mexico COPCORD methodology	Mexico	Adult identified as Raramuri indigenous resident in Chihuahua city, Mexico	MSK pain and rheumatic diseases were highly prevalent, with high impact on daily activities
Granados et al., 2016 [26]	Estimate the prevalence of MSK disorders and rheumatic diseases in the Warao, Kari'ña and Chaima indigenous populations (Monagas State) Venezuela COPCORD methodology	Venezuela	Adult identified as Warao, Kari'ña and Chaima indigenous in Monagas State, Venezuela	The prevalence of MSK disorders and rheumatic diseases was high in the three indigenous groups. The Chaima indigenous group reported a higher prevalence of rheumatic diseases
Alvarez-Nemegyei et al., 2016 [27]	Estimate the prevalence of RRPS in four Latin-American indigenous groups COPCORD methodology	Mexico, and Argentina	Adult identified as in four indigenous groups: Chontal (Oaxaca, Mexico), Mixteco (Oaxaca, Mexico), Maya-Yucateco (Yucatán, Mexico) and Qom (Argentina)	There was a consistently higher prevalence of RRPS in indigenous populations RRPS vary according to the population and may be influenced by inherent factors specific for each population

US, United States; UK, United Kingdom; MSK, rheumatic musculoskeletal disorders; MSDs, work-related musculoskeletal disorders; COPCORD, Community-Oriented Program for Control of Rheumatic Diseases; CTS, carpal tunnel syndrome; OA, osteoarthritis; SGIx, Social Gap Index; RRPS, rheumatic regional pain syndromes.

Table 1. Summary and analysis of quantitative studies.

In 2004, Vindigni et al. [14] describes the prevalence of musculoskeletal (MSK) disorders in rural Indigenous Australians areas, where the 57% of people have learnt to live with chronic levels of pain affecting multiple anatomical sites.

Ahonen et al. [16] summarize the information on immigrant occupational health available from studies, incorporating varied study designs. The immigration was associated with occupational risk factors, health consequences, and the social, economic and cultural influences on worker health. Similarly, Jørgensen et al. [7] investigate differences in self-reported health measures between immigrant and Danish cleaners, where the immigrant cleaners generally had a poorer self-reported health and work ability than the Danish cleaners.

Joshi and Chopra [18] and Davatchi et al. [19] investigate the prevalence of MSK diseases in India and Iran, respectively, through the implementation of the Community Orientated

Program for the Control of Rheumatic Diseases (COPCORD). In the first case, people living in urban area had less prevalence of MSK diseases, in comparison to the people in rural area. In the second case, the musculoskeletal complaints were more frequent in Turks than in Caucasians, both living in identical environment.

In United States (US), several papers have been published on this topic, especially on immigrant population. In this sense, Schulz et al. [5] describes that the back symptoms and wrist/hand symptoms were reported by over 35% of Latino workers. In the same way, Rosenbaum et al. [6] identifies that upper body musculoskeletal and low back pain are common in immigrant Latino workers and may negatively impact long-term health and contribute to occupational health disparities. Xiao et al. [8] and Cartwright et al. [9] describe the Latino population as a vulnerable group for MSDs.

In Latin American (LA), previous studies have shown that people belonging to an indigenous group are associated with the prevalence of rheumatic disease as the rheumatoid arthritis (RA) [28]. Due to the presence in LA of indigenous groups, all countries and their condition of vulnerability, not only socioeconomic but also ethnic, were created the Latin American Group for the Study of Rheumatic Diseases in Original Populations (Grupo Latino Americano De estudio de Enfermedades Reumáticas en Pueblos Originarios; GLADERPO). The main objective was to carry out epidemiological, anthropological and genetic studies, thus fulfilling with intervention processes in the affected populations. Studies are currently being carried out in several indigenous populations of Argentina, Mexico and Venezuela [29]. In 2016, "The Clinical Rheumatology Journal" published a supplement with information about the Maya-Yucateco, Mixtec, Chontal and Rarámuri populations from Mexico; Warao, Kari'ña and Chaima from Venezuela; and Qom from Argentina [29]. The design of the studies, including samplings, case definition and methodology, was the same. The methodology was that proposed by COPCORD. Overall, low back pain, osteoarthritis (OA) and rheumatic regional pain syndromes (RPPS) were the most prevalent rheumatic diseases across all populations. Among the inflammatory rheumatic diseases, RA was the most prevalent, especially in the Qom community (2.4%) [22]. There were variations in the prevalence of certain diseases among different populations. The low back pain was more prevalent in the Qom community (19.8%) [22] and the OA in the Chontal community (32.1%) [24]. Variations were not related to the design of the study but to the characteristics of the populations and environmental factors, that is to say, heavy loads in Chontales versus Rarámuris or Maya-Yucateco in people not exposed to the same environment [20–23, 26, 27]. These results give information on working and socioeconomic conditions in these populations.

2. Synthesis and interpretation of qualitative studies: Few researches have been conducted based on social impacts associated with MSDs so far. Most outcome studies in occupational health have been focused on workers' compensation (WC), insurance payments, provision of medical services, return-to-work time and other direct insurance and employment-related measures [30]. Unfortunately, there has been little research studying the impact on the social and family environment of the workers affected, as well as the indirect economic consequences. These complex interactions create significant difficulties for researchers attempting to study the social consequences of MSDs.

http://dx.doi.org/10.5772/67002

The sociodemographic characteristics of affected individuals and groups such as age, gender, ethnicity, nationality, education and socioeconomic status can influence the social consequences of the injury. They could also influence the responses of employers, insurers and medical providers [30].

Studies suggest that work-related injuries have significant long-term physical, economic and psychological consequences, which were worse in those who had been out of work for longer periods of time [31].

Patients with more serious MSDs have higher rates of psychological problems, drug abuse, and marital difficulties and the quality of the affected worker's family relationships.

Contemporary studies suggest that significant disparities in the incidence of MSDs and deaths still exist among various racial and ethnic groups. For example, in California, workplace injuries were found to occur 32% more frequently among black workers than among whites, and the rate for Hispanics is 18% higher [32]. There is scattered evidence indicating that the social consequences of MSDs also fall most heavily on women, minorities, immigrants and other vulnerable populations [33, 34].

4. Discussion

It is extremely important to study in-depth the closely related social and cultural aspects. Only traditional social variables have been studied that are only the tip of the iceberg. Multilevel analysis (e.g., individuals, social environment, health system, employers, type of work, insurance) should be incorporated for study between MSDs and vulnerable groups [35].

The constant migration of vulnerable groups to developed countries and precarious work must be taken into account in the analysis of MSDs.

The approach to this problem should be through the qualitative methods. These can serve to contextualize quantitative data providing means of cross-validation and what is termed by social scientists as triangulation, that is, the use of different approaches, by the conceptual, methodological or data collecting, to study the same problem in order to optimize the understanding of underlying mechanisms, work activity and environment, relationships, and solutions.

The incorporation of the COPCORD methodology should be of great help because it would assist the health system in the appropriate selection and application of resources, as well as the decision making in the system, demonstrating that the major worldwide problems of rheumatic complaints and disability are not just relevant to the elderly populations of developed countries, but also to the vulnerable populations living in poorer conditions [36].

5. Conclusion and recommendations

The burden of MSK disorders is likely to vary in different parts of the world. While modernization has bulldozed the Western world toward similar lifestyles, it has not yet transformed the cultural and traditional picture of Asia, Africa and several regions of South America. The disease health process is cultural and is influenced by socioeconomic factors. Many indigenous groups traditionally squat and/or sit cross-legged on the ground to enhance their daily activities [37, 38]. Several MSDs can interfere causing immense suffering and frustration. Despite severe pain and disability, people do not easily give up this traditional and cultural lifestyle. It also coping to pain and disability, giving priority to work. This can also be seen in different vulnerable groups, where the quality and precariousness of work make individuals normalize pain and coping to it.

The recommendations are based on the literature discussed in this chapter. The assistance to this problem should be multidisciplinary, using both qualitative and quantitative methodology.

A priority is the promotion and prevention of such conditions in the most vulnerable population groups, living in precarious conditions and social inequality.

Acknowledgements

We thank Gabriel Crocce for assistance in translation.

Conflict of interest

None.

Author details

Rosana M. Quintana

Address all correspondence to: rosanaquintana@gmail.com

Rheumatology Service, Provincial Hospital of Rosario, Rosario, Argentina

References

[1] Punnett L, Wegman DH. Work-related musculoskeletal disorders: the epidemiologic evidence and the debate. J Electromyogr Kinesiol. 2004 Feb;14(1):13–23.

[2] Rempel DM, Punnett L. Epidemiology of wrist and hand disorders, In: M. Nordin, G.B. Andersson, M.H. Pope (Eds.), Mosby-Year Book, Inc, Philadelphia, PA, 1997, pp. 421–430.

[3] WHO. Identification and control of work related diseases. Technical Report Series no. 714, World Health Organisation, Geneva, Switzerland, 1985.

[4] Eltayeb SM, Staal JB, Hassan AA, Awad SS, de Bie RA. Complaints of the arm, neck and shoulder among computer office workers in Sudan: a prevalence study with validation of an Arabic risk factors questionnaire. Environ Health. 2008 Jun 27;7:33. doi:10.1186/1476-069X-7-33

[5] Schulz MR, Grzywacz JG, Chen H, Mora DC, Arcury TA, Marín AJ, Mirabelli MC, Quandt SA. Upper body musculoskeletal symptoms of Latino poultry processing workers and a comparison group of Latino manual workers. Am J Ind Med. 2013 Feb;56(2):197–205. doi:10.1002/ajim.22100. Epub 2012 Jul 27.

[6] Rosenbaum DA, Grzywacz JG, Chen H, Arcury TA, Schulz MR, Blocker JN, Mora DC, Quandt SA. Prevalence of epicondylitis, rotator cuff syndrome, and low back pain in Latino poultry workers and manual laborers. Am J Ind Med. 2013 Feb;56(2):226–234. doi:10.1002/ajim.22127. Epub 2012 Oct 25.

[7] Jørgensen MB, Rasmussen CD, Carneiro IG, Flyvholm MA, Olesen K, Ekner D, Søgaard K, Holtermann A. Health disparities between immigrant and Danish cleaners. Int Arch Occup Environ Health. 2011 Aug;84(6):665–674. doi:10.1007/s00420-010-0607-2. Epub 2011 Jan 4.

[8] Xiao H, McCurdy SA, Stoecklin-Marois MT, Li CS, Schenker MB. Agricultural work and chronic musculoskeletal pain among Latino farm workers: the MICASA study. Am J Ind Med. 2013 Feb;56(2):216–225. doi:10.1002/ajim.22118. Epub 2012 Sep 28.

[9] Cartwright MS, Walker FO, Newman JC, Schulz MR, Arcury TA, Grzywacz JG, Mora DC, Chen H, Eaton B, Quandt SA. One-year incidence of carpal tunnel syndrome in Latino poultry processing workers and other Latino manual workers. Am J Ind Med. 2014 Mar;57(3):362–369. doi:10.1002/ajim.22250. Epub 2013 Aug 31.

[10] Greenhalgh T, Wong G, Westhorp G, Pawson R. Protocol--realist and meta-narrative evidence synthesis: evolving standards (RAMESES). BMC Med Res Methodol. 2011 Aug 16;11:115. doi:10.1186/1471-2288-11-115

[11] Urwin M, Symmons D, Allison T, Brammah T, Busby H, Roxby M, Simmons A, Williams G. Estimating the burden of musculoskeletal disorders in the community: the comparative prevalence of symptoms at different anatomical sites, and the relation to social deprivation. Ann Rheum Dis. 1998 Nov;57(11):649–655.

[12] Mergler D. Combining quantitative and qualitative approaches in occupational health for a better understanding of the impact of work-related disorders. Scand J Work Environ Health. 1999;25(Suppl 4):54–60.

[13] Cole DC, Ibrahim SA, Shannon HS, Scott F, Eyles J. Work correlates of back problems and activity restriction due to musculoskeletal disorders in the Canadian national population health survey (NPHS) 1994-5 data. Occup Environ Med. 2001 Nov;58(11):728–734.

[14] Vindigni D, Griffen D, Perkins J, Da Costa C, Parkinson L. Prevalence of musculoskeletal conditions, associated pain and disability and the barriers to managing these conditions in a rural, Australian Aboriginal community. Rural Remote Health. 2004 Jul–Sep;4(3):230. Epub 2004 Aug 27.

[15] Turner JA, Franklin G, Fulton-Kehoe D, Egan K, Wickizer TM, Lymp JF, Sheppard L, Kaufman JD. Prediction of chronic disability in work-related musculoskeletal disorders: a prospective, population-based study. BMC Musculoskelet Disord. 2004 May 24;5:14.

[16] Ahonen EQ, Benavides FG, Benach J. Immigrant populations, work and health--a systematic literature review. Scand J Work Environ Health. 2007 Apr;33(2):96–104.

[17] Côté P, van der Velde G, Cassidy JD, Carroll LJ, Hogg-Johnson S, Holm LW, Carragee EJ, Haldeman S, Nordin M, Hurwitz EL, Guzman J, Peloso PM. The burden and determinants of neck pain in workers: results of the Bone and Joint Decade 2000-2010 Task Force on Neck Pain and Its Associated Disorders. J Manipulative Physiol Ther. 2009 Feb;32(2 Suppl):S70–S86. doi:10.1016/j.jmpt.2008.11.012

[18] Joshi VL, Chopra A. Is there an urban-rural divide? Population surveys of rheumatic musculoskeletal disorders in the Pune region of India using the COPCORD Bhigwan model. J Rheumatol. 2009 Mar;36(3):614–622. doi:10.3899/jrheum.080675. Epub 2009 Feb 4.

[19] Davatchi F, Jamshidi AR, Tehrani Banihashemi A, Gholami J, Hossein Forouzanfar M, Akhlaghi M, Barghamdi M, Noorolahzadeh E, Khabazi AR, Salesi M, Salari AH, Karimifar M, Essalat-Manesh K, Hajialiloo M, Soroosh M, Farzad F, Moussavi HR, Samadi F, Ghaznavi K, Asgharifard H, Zangiabadi AH, Shahram F, Nadji A, Akbarian M, Gharibdoost F, Rasker JJ. Effect of ethnic origin (Caucasians versus Turks) on the prevalence of rheumatic diseases: a WHO-ILAR COPCORD urban study in Iran. Clin Rheumatol. 2009 Nov;28(11):1275–82. doi:10.1007/s10067-009-1235-7. Epub 2009 Jul 26.

[20] Peláez-Ballestas I, Alvarez-Nemegyei J, Loyola-Sánchez A, Escudero ML. Prevalence and factors associated with musculoskeletal disorders and rheumatic diseases in indigenous Maya-Yucateco people: a cross-sectional community-based study. Clin Rheumatol. 2016 Jul;35(Suppl 1):15–23. doi:10.1007/s10067-015-3085-9. Epub 2015 Oct 5.

[21] Rodriguez-Amado J, Moreno-Montoya J, Alvarez-Nemegyei J, Goycochea-Robles MV, Sanin LH, Burgos-Vargas R, Cardiel MH, Garza-Elizondo MA, Maradiaga M, Pelaez-Ballestas I; GEEMA. The Social Gap Index and the prevalence of osteoarthritis in the community: a cross-sectional multilevel study in Mexico. Clin Rheumatol. 2016 Jan;35(1):175–182. doi:10.1007/s10067-014-2776-y. Epub 2014 Sep 18.

[22] Quintana R, Silvestre AM, Goñi M, García V, Mathern N, Jorfen M, Miljevic J, Dhair D, Laithe M, Conti S, Midauar F, Martin MC, Barrios MC, Nieto R, Prigione C, Sanabria A, Gervasoni V, Grabbe E, Gontero R, Peláez-Ballestas I, Pons-Estel BA. Prevalence of musculoskeletal disorders and rheumatic diseases in the indigenous Qom population of Rosario, Argentina. Clin Rheumatol. 2016 Jul;35(Suppl 1):5–14. doi:10.1007/s10067-016-3192-2. Epub 2016 Feb 6.

[23] Loyola-Sanchez A, Richardson J, Pelaez-Ballestas I, Alvarez-Nemegyei J, Lavis JN, Wilson MG, Wilkins S. The impact of arthritis on the physical function of a rural Maya-Yucateco community and factors associated with its prevalence: a cross sectional, community-based study. Clin Rheumatol. 2016 Jul;35(Suppl 1):25–34. doi:10.1007/s10067-015-3084-x.

[24] Julián-Santiago F, García-García C, García-Olivera I, Goycochea-Robles MV, Pelaez-Ballestas I. Epidemiology of rheumatic diseases in Mixtec and Chontal indigenous communities in Mexico: a cross-sectional community-based study. Clin Rheumatol. 2016 Jul;35(Suppl 1):35–42. doi:10.1007/s10067-015-3148-y

[25] Del Río Nájera D, Santana N, Peláez-Ballestas I, González-Chávez SA, Quiñonez-Flores CM, Pacheco-Tena C. Prevalence of rheumatic diseases in Raramuri people in Chihuahua, Mexico: a community-based study. Clin Rheumatol. 2016 Jul;35(Suppl 1):43–52. doi:10.1007/s10067-016-3225-x

[26] Granados Y, Rosillo C, Cedeño L, Martínez Y, Sánchez G, López G, Pérez F, Martínez D, Maestre G, Berbin S, Chacón R, Stekman I, Valls E, Peláez-Ballestas I. Prevalence of musculoskeletal disorders and rheumatic disease in the Warao, Kari'ña, and Chaima indigenous populations of Monagas State, Venezuela. Clin Rheumatol. 2016 Jul;35(Suppl 1):53–61. doi:10.1007/s10067-016-3194-0

[27] Alvarez-Nemegyei J, Peláez-Ballestas I, Goñi M, Julián-Santiago F, García-García C, Quintana R, Silvestre AM, García-Olivera I, Mathern NA, Loyola-Sanchez A, Conti S, Sanabria AJ, Pons-Estel BA. Prevalence of rheumatic regional pain syndromes in Latin-American indigenous groups: a census study based on COPCORD methodology and syndrome-specific diagnostic criteria. Clin Rheumatol. 2016 Jul;35(Suppl 1):63–70. doi:10.1007/s10067-016-3188-y

[28] Moreno-Montoya J, Alvarez-Nemegyei J, Sanin LH, Pérez-Barbosa L, Trejo-Valdivia B, Santana N. Association of regional and cultural factors with the prevalence of rheumatoid arthritis in the Mexican population: a multilevel analysis. JCR. 2015;21:57–62. doi:10.1097/RHU.0000000000000223

[29] Peláez-Ballestas I, Pons-Estel BA, Burgos-Vargas R. Epidemiology of rheumatic diseases in indigenous populations in Latin-Americans. Clin Rheumatol. 2016 Jul;35(Suppl 1):1–3. doi:10.1007/s10067-016-3298-6. Epub 2016 May 14.

[30] Dembe AE. The social consequences of occupational injuries and illnesses. Am J Ind Med. 2001 Oct;40(4):403–417.

[31] Pransky G, Benjamin K, Hill-Fatouhi C, Himmelstein J, Fletcher K, Katz J, Johnson W. 2000. Outcomes in work-related upper extremity and low back injuries: results of a retrospective study. Am J Ind Med. 2000;37(4):400–9.

[32] Robinson JC. Trends in racial inequality and exposure to work-related hazards, 1968-1986. Milbank Q. 1987;65(Suppl. 2):404–20.

[33] Lerner D, Allaire SH, Reisine ST. Work disability resulting from chronic health conditions. J Occup Environ Med. 2005 Mar;47(3):253–264.

[34] Taylor W, Smeets L, Hall J, McPherson K. The burden of rheumatic disorders in general practice: consultation rates for rheumatic disease and the relationship to age, ethnicity, and small-area deprivation. N Z Med J. 2004 Oct 8;117(1203):U1098.

[35] Pelaez I, Infante C, Quintana R. Help-seeking trajectory in patients with rheumatoid arthritis. Clin Rheumatol. 2015 Mar;34(Suppl 1):S17–S28. doi:10.1007/s10067-015-3013-z. Epub 2015 Jul 23.

[36] Muirden KD. Community Oriented Program for the Control of Rheumatic Diseases: studies of rheumatic diseases in the developing world. Curr Opin Rheumatol. 2005 Mar;17(2):153–156.

[37] Chopra A, Abdel-Nasser A. Epidemiology of rheumatic musculoskeletal disorders in the developing world. Best Pract Res Clin Rheumatol. 2008 Aug;22(4):583–604. doi:10.1016/j.berh.2008.07.001

[38] Palazzo C, Ravaud JF, Trinquart L, Dalichampt M, Ravaud P, Poiraudeau S. Respective contribution of chronic conditions to disability in France: results from the national Disability-Health Survey. PLoS One. 2012;7(9):e44994. Epub 2012 Sep 14.

Section 4

Occupational Protection

Chapter 11

Working in Cold Environment: Clothing and Thermophysiological Comfort

Radostina A. Angelova

Additional information is available at the end of the chapter

http://dx.doi.org/10.5772/65687

Abstract

The chapter presents an in-depth discussion over the occupational activities in a cold environment, which can be performed both outdoors and indoors. It explores the differences between working in natural and artificial cold environment. The thermophysiological comfort, the reactions of the thermoregulatory system during cold exposure, and cold-related injuries are presented and discussed in detail. Clothing as the only insulating barrier between the body and the cold environment is discussed, and hi-tech solutions for development of cold protective clothing are presented. The particular application of standards for the indoor environment is considered, and their input for the proper management of the occupational activities in the cold is analyzed.

Keywords: cold environment, thermophysiological comfort, protective clothing, cold-related injuries

1. Introduction

Being warm-blooded creatures, humans can maintain a stable temperature of their bodies despite the large temperature fluctuations of the surrounded environment. The thermophysiological comfort is one of the aspects of the human comfort both indoors and outdoors. It is preconditioned by the functioning of the thermoregulatory system of the body and its reactions to the temperature of the surrounding air, the activity performed, and the used clothing insulation. Body's thermoregulation is very sophisticated and tries to maintain the temperature of the core body around 37°C while balancing between the heat, gained or produced, and the heat, released by the body to the surroundings.

According the definition in [1], cold is any environment, where people are exposed to a temperature below 15°C. The British standard BS7915 (1998) [2] determines that "cold environment" is the one with an air temperature below 12°C. People are working in a risky cold environment, which can be natural (outdoor) or artificial (indoor) cold environment. It is the clothing only (in the past) and special protective clothing (at present) that is the only barrier between the human body and the cold environment. Clothing, activity, and proper management of the occupational activities help and maintain the thermophysiological comfort of people, working in the cold.

The purpose of this chapter is to present details about the human thermoregulatory system and the thermophysiological reactions of the body in a cold environment. Natural cold environment and artificial cold environment are compared to the light of their different effects on the occupational activities and management of the cold exposure. Cold-related injuries as part of the thermoregulatory reactions of the body and the risk from cold exposure are summarized. Clothing and hi-tech garments for protection from extreme temperatures are discussed. Practical advices and standards for the management of the occupational activities in the cold are presented.

2. Thermophysiological comfort and thermoregulation

2.1. Thermophysiological comfort

Comfort is a relative and subjective category, which depends on individual reactions and perceptions. Hatch [3] defines the comfort as a neutral state, where there is no pain or discomfort. The *thermophysiological comfort* is a part of the *physiological comfort*, related to the reactions of the thermoregulatory system [4]. It is based on the sensors for warmth and cold in the body, susceptible to thermal environment, air velocity, temperature asymmetry, etc. and the consecutive reactions of the thermoregulatory system, which increases either the heat, produced by the body, or the heat losses to the surrounding environment. When the heat production is equal to the heat losses, the body is considered to be in a state of a thermophysiological comfort [4].

2.2. Basics of human thermoregulation in the cold

The main goal of the thermoregulatory system is to maintain the temperature of the core body (the brain and organs in the torso) around 37°C. The sensation of cold is initiated by the reaction of cold receptors: specialized nerve endings in the skin. They send signals to the central nervous system together with signals, coming from the brain (**Figure 1**). The two signals are processed in the hypothalamus: the gland in the brain that is responsible for the reactions of the thermoregulatory system. Actually, the spinal cord also controls the thermoregulation in case of cold exposure [5].

When the hypothalamus is activated, it sends electrical signals that trigger different thermoregulatory mechanisms, related to the decrement of the heat losses from the body (skin and lungs) to the surrounding air and increment of the heat production (in muscles and liver) [6–10]. **Figure 2** summarizes the basic physical and physiological reactions of the human body to the cold.

One of the first reactions of the body in cold environment is to provoke vasoconstriction: a decrement of the cross section of the blood vessels in the surface zones of the skin and the

http://dx.doi.org/10.5772/65687

extremities. The vasoconstriction appears due to signals from the hypothalamus to the smooth muscles in the arterioles. The aim of the vasoconstriction is to impede the transfer of hot blood out of the core body, thus preserving as long as possible the accurate functioning of the heart, brain, and lungs. This process can increase the body temperature of 1–2°C, due to the reduction of heat losses through the three mechanisms of heat transfer: conduction, convection, and radiation.

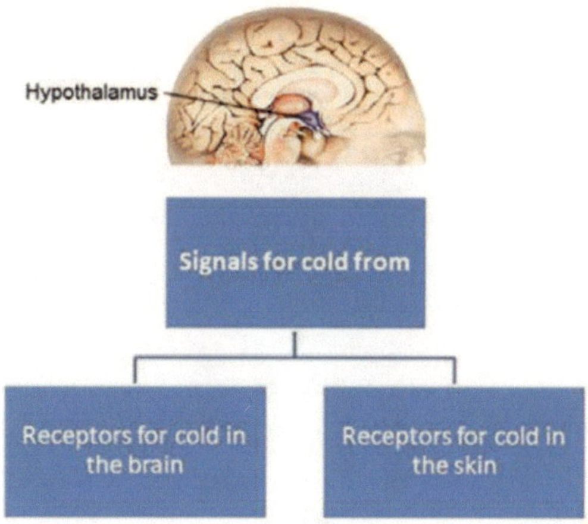

Figure 1. Signals sent to the hypothalamus.

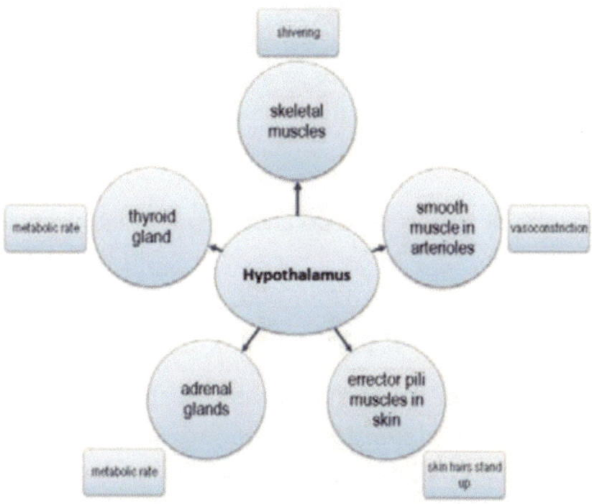

Figure 2. Thermoregulatory mechanisms in cold environment.

Figure 3 presents thermograms of a human hand, before and 3 min after the exposure to a cold environment of −10°C. The isolines clearly show the vasoconstriction process that has started. Vasoconstriction creates a feeling of cold, causing muscle tremors and increased heat production. Heat production from the liver also adds to the process. Cold hands and feet are the most frequent complaints of people in low ambient temperatures. Many of cold-related injuries are referred to fingers, nose, ears, and extremities, due to the decreased blood flow to them.

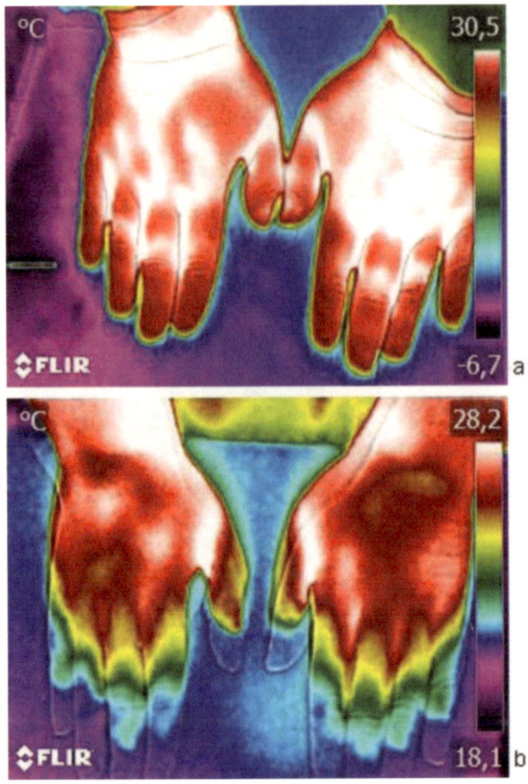

Figure 3. Thermograms of human hand: (a) before the cold exposure and (b) after a cold exposure to −10°C.

The behavioral reaction of the person to add an additional layer(s) of clothing (or bedding) also contributes to the augment of the core body temperature and the reduction of heat losses from the skin. Thus, when exposed to cold environment, the body maintains its internal temperature via vasoconstriction, increased heat production, and behavioral changes.

Signals to the erector pili muscles in the skin provoke pricking of the skin. Though small as an effect, the reaction is related to the detention of motionless air as close to the skin surface as possible. The air near the skin can be an additional insulation layer for the clothed body, as the thermal resistance of the air is almost twice higher than the thermal resistance of the natural fibers, for example, used in textile and clothing production.

The skeletal muscles are triggered to shivering: a spontaneous movement of the muscles, aiming to increase the heat production. This reaction to the cold is temporary and depends on the glycogen, the "fuel" of the muscles. The glycogen depletion stops the heat production by shivering.

The signals from the hypothalamus to the thyroid and adrenal glands have to increase the metabolism, which is another source of heat production for the body. The metabolic reactions are performed on a cell level.

http://dx.doi.org/10.5772/65687

2.3. Body reactions to cold

Despite its complex and sensitive mechanism for thermoregulation, the human body is coping with the aggressive impact of the environment in a relatively narrow range. The decrement of the core body temperature has an adverse effect on the function of the body and can lead to severe disability and even death.

Out of these threats, however, even in cases, when the protective mechanisms of the body are sufficiently effective and thermoregulatory processes are able to maintain the core body temperature at about 37°C, the cold environment can cause a sequence of events, associated with the so-called thermal discomfort (**Figure 4**). The low environmental temperature causes delay in reaction time, fatigue, and increased sleepiness of the individual [11].

These effects are, to their nature, pure physiological processes, i.e., the accelerated heart rate causes fatigue, and the lack of irrigation in the brain tissue causes sleepiness. All these can result in a larger number of subjective errors, compared to the occupational activities in thermal comfort environment (20–22°C).

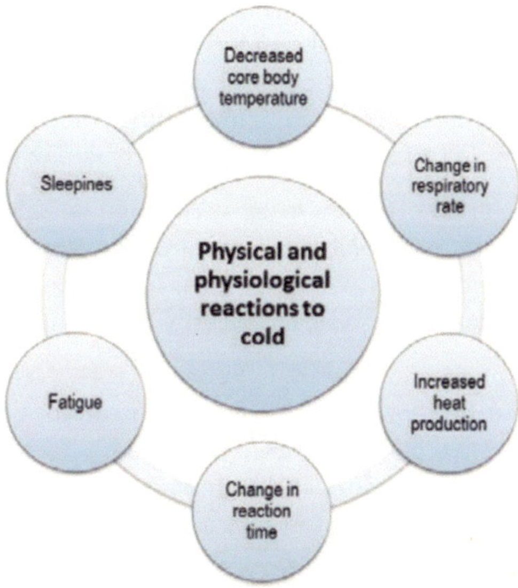

Figure 4. Body's reactions to cold environment.

3. Particularities of working in cold environments: natural vs. artificial cold

The work in natural and artificial cold has very similar features: in both cases the impact of the low and sub-zero temperatures could be dangerous for the human body. However, the majority of the workplaces in the cold are related to an outdoor exposure [12]. In winter time, in mountain, arctic and subarctic regions, different logistic activities for the society (road maintenance, transport) involve cold exposure. Tourism and winter sports also are associated with occupational activities that require cold exposure. Different industries like construction industry, fishery, farming, reindeer breeding, mining, metallurgy, forestry, horticulture, etc. involve work in outdoor conditions with low temperatures. In Sweden over 30% of the employed persons, for example, work in cold conditions repeatedly, for shorter or longer time [13].

The artificial cold workplaces can be found mainly in the food industry and fishery. Fresh food is usually preserved at a temperature below 6°C; frozen food is handled and stored at a temperature around −25°C. The work in food processing departments of big supermarkets is also related to cold exposure. It is reported that cold-related diseases and discomfort complaints are more frequently observed among workers, exposed to artificial than to outdoor cold [14, 15].

In any case, to protect the human body from the cold-related hazards, workers must be provided with protective clothing during their cold exposure. The comparison between the occupational activities in natural and artificial cold shows, however, some differences, which influence the thermophysiological comfort of the workers.

On the first place, the indoor, artificial cold, is more stable in temperature, air velocity, and humidity fluctuations, which is preconditioned by the application of systems for chilling and air conditioning. This helps the proper selection of protective clothing, which does not need changing during the exposure. In the natural cold environment, the air temperature changes during the 24-hour period. This requires the use of clothing, which allows adding or removing of layers within some limits, in order to protect the worker from overheating or freezing.

In addition, the climate conditions have to be considered together with the geographical features: activities in mountain regions (road construction, logging, tourism) increase the negative effect of the low temperatures and the solar load; activities in the flatlands or seas (agriculture, fishing, oil platforms) increase the severity of the wind effect.

In the food processing industry, static work is frequently observed. Some outdoor activities (transportation or work with heavy machinery) also require a sitting posture of the worker, but it is performed in the protective indoor environment of the vehicle (track, bulldozer, etc.). In any case the immobility of the body in cold environment is harmful and has to be avoided. Protective gloves and boots must be used, but they cannot replace the need of blood circulation in the extremities. At the same time, indoor cold work may require fine motor skills activities, which is in contradiction with the application of heavy protective gloves and mittens.

Another difference between the occupational activities in artificial and natural cold is that workers in an artificial cold environment move more frequently between colder and warmer environment. The temperature difference provokes higher strain on the thermoregulatory system of the body. At the same time, continuous cold air flows in the artificial cold facilities lead to appearance of body temperature asymmetry (asymmetric cooling), which adds to the thermal discomfort of the workers.

Last but not least, the protective clothing in an artificial cold environment may need to combine different protective abilities, i.e., against chemical or mechanical hazards, because of the occupational safety regulations.

4. Cold-related injuries

Cold-related injuries may occur in any environment, which temperature is around the freezing temperature. However, the thermoregulatory reactions to cold start at any air temperature, which is below the temperature of the body, and signals for cold are received by the cold sensors in the skin. Hypothermia may occur even in deserts if it is preconditioned by the body state and environmental conditions.

The major injuries, related to cold exposure, are frostbite, trench foot, immersion foot, hypothermia, and cold allergy. The reactions of the thermoregulatory system against the cold start when the core body temperature decreases below 37°C (**Figure 5**). The shivering, which is a way for heat production, reaches its peak at 35°C [16]. Below this temperature the body starts to demonstrate signs of hypothermia. At a temperature of 34°C, muscle rigidity appears, and the person is not anymore able to perform manual operations due to a loss of manual dexterity [17]. At 32°C the consciousness is clouded and around 30°C it is lost. If the core body temperature decreases below 30°C, the risk of death becomes extremely high if the cold exposure continues. The data from [16] show that cardiac arrhythmia appears at 29°C, and at 27°C the person appears dead. The drop of the core body temperature to 24°C leads to the development of pulmonary edema, and around 20°C the heart stops beating. However, it has to be mentioned that the lowest core body temperature, from which a person has been recovered, is 18°C [16].

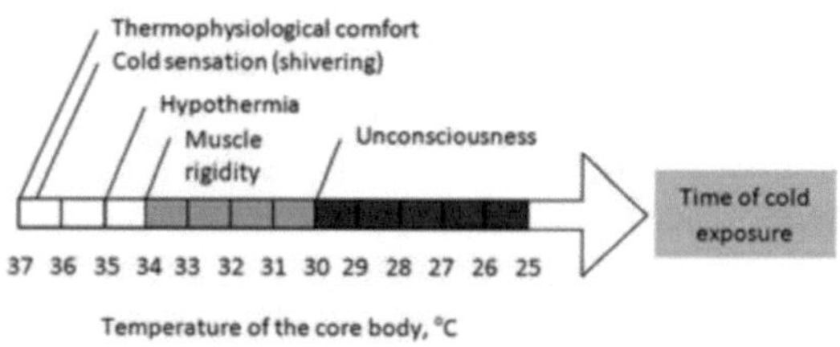

Figure 5. Effect of the time of cold exposure on core body temperature and hypothermia.

Frostbite is an injury that affects mainly the extremities in the cold, especially fingers and toes, as well as unprotected parts of the face (nose, lips, ears). It appears due to the crystallization of the liquids in the cells of the skin and deeper tissues of the body shell. The severity of the injury depends on the speed of freezing. Temperatures below 0°C and low relative humidity of the air precondition the appearance of frostbite.

Trench foot and immersion foot are classified as a nonfreezing cold injury [16]. Trench foot may appear at temperatures above 0°C and high relative humidity of the air. Its severity depends on the time of cold exposure. Immersion foot occurs as a result of prolonged static work in upright posture or as a consequence of immobilization of the extremities. It is related with wet environment, including cold water exposure.

Hypothermia appears after prolonged cold stimulus, but other environmental factors like wind, contact with cold objects, or immersion in water can favor it. Three types of hypothermia are classified in [18]: primary, secondary, and clinically induced (the last being out of the scope of cold-related injuries).

Primary hypothermia is diagnosed when the thermoregulation responses to the cold exist, but they cannot beat the symptoms of hypothermia. Primary hypothermia can appear in any cold environment, when the low temperature overwhelms the body thermoregulatory system.

Secondary hypothermia differs from the primary as the thermoregulatory reactions of the body are impaired [17]. Symptoms of hypothermia appear again, but their severity is not proportional to the cold induced. Secondary hypothermia is due to additional peculiarities of human physical or physiological state: fatigue, illness, or injury. Insufficient clothing insulation, poor nutrition or dehydration, and short sleep are also reasons for secondary hypothermia. The onset of the secondary hypothermia can hardly be predicted, being dependent of several factors: clothing insulation, metabolic rate, body size, nutrition status, hydration, and physiological and even psychological status.

Cold allergy, expressed in red and itchy pimples on uncovered skin, exposed to cold air, is also frequently observed as cold injury. It usually disappears after warming. Severe cases of cold allergy are associated with fever, seizures, fever, increment of the heartbeat, and swelling of the torso or extremities.

5. Textiles and clothing for protection from cold

Clothing plays the role of a passive insulation layer between the body and the environment. Both the textile layers and the layers of air between them in the clothing system perform this role. The adjustment of the passive insulation to the changing thermal conditions of the environment may be done through adding or removing textile layers. However, sometimes this is not possible to be done, or the textile layers, necessary to guarantee the body thermal comfort, are so thick that the movements will be limited.

The textiles, in the form of layers in clothing garments, headwear, handwear, or footwear, are of crucial importance for the survival and healthy work of people in the cold environment, as they are the only barrier between the human body and the cold. The ability of the textile layers to transfer the moisture from the body to the environment, while keeping the heat next to the skin and preventing the moisture transfer from the surroundings to the body, determines to a great extent the thermophysiological comfort of humans in the cold [19].

The main aim of the cold protective clothing is to entrap as much as possible air between the textile layers; at the same time, the clothing ensemble has to be as light as possible to assure proper activities and unimpeded movements. Therefore, different layers of clothing have to be worn—besides the better air encapsulation, the number of layers gives the advantage to remove one or more of them if the body's heat production increases (during activity) or the solar radiation heats the body in outdoor conditions.

At least three layers of loose-fitting clothing must be worn [19]. The inner, closer to the body layer, is usually made of polyester, polypropylene, or other synthetic fibers that draw moisture

http://dx.doi.org/10.5772/65687

away from the skin and keep it dry. The middle layer is the most insulating one; it is made of down, wool, nonwoven webs of synthetic fibers, etc. and holds the body's heat. The outer layer aims to protect the body from wind and precipitation; it is made of "breathable" waterproof fabrics that allow some ventilation (like Gore-Tex® or polyamide). This layer may frequently need to be resistant to oil, fire, chemicals, or abrasion.

The fitting of each clothing item in the ensemble is extremely important. The tight clothes may press the body tissues and decrease the blood flow, thus increasing the risk of cold injuries. Any additional layer(s) of clothing should be large enough not to compress the inner layers and decrease the insulation properties due to omitting the insulating air layer between two consecutive textile layers.

Proper selection of footwear and handwear has to be done to prevent hands and feet from the cold. Mittens are better solution than gloves, as four of the fingers share one and the same "thermal environment," but gloves are needed if hands' finer movements are a key to the performed activity in the cold. The head has to be obligatory protected by a cap, as the heat may be seriously lost through the head to the environment when the other parts of the body are well insulated. The chicks and nose may be protected from the cold by a mask. In case of obligatory use of a helmet, a wool knit cap has to be worn beneath.

Insulated boots with removable felt liners are mostly used for protection in the cold. The liners and the socks are an important part of the footwear. They have to be kept dry so as to perform best in the cold environment as an insulating layer. Inner socks, made of polypropylene that helps keep feet dry and warm, are best to be combined with outer thicker socks. The boot liners have to be removed daily for complete drying. The socks have to be also changed if they get wet or damp.

The incorporation of phase change materials (PCMs) in textiles for cold protection adds an active thermal insulation effect to the passive insulation, performed by the clothing [20]. PCMs, applied in cold protective clothing, are mostly paraffins from different types, used in several combinations. PCMs can experience the process of change from one state to another, i.e., from liquid to solid and back; thus, they are able to absorb, store, or discharge heat, following the fluctuations of the surrounding temperature.

The paraffin is incorporated into microcapsules, which has to prevent its dissolution during the liquid phase. Outlast Technologies Inc., which is the leader in production of textiles with PCMs, have developed Thermocules microcapsules, which can be added in the structure of the synthetic fibers in the wet spinning process or can be incorporated in flat textiles (woven, knitted, and nonwoven) as part of surface coating.

Classical cold wear garments can hardly ensure the thermophysiological comfort of the body in extreme cold conditions for a long time. The application of PCMs can help the process of avoiding thermal stress and hypothermia during occupational activities, thus increasing the work performance under high thermal stress.

Heated clothing and accessories are an alternative of the classical cold protective clothing. They are wired and use different technologies to provide warmth to the body parts that need it with priority: core body, feet, and hands.

The Microwire™ technology of Gerbing uses thin conductive filaments, coated with Teflon® that form heating panels, incorporated in clothing items. In jackets the panels are placed on the back, chests, and collar. Incorporated in gloves, the Microwire™ heating panels help sustaining the finer finger movements as long as possible. PrimaLoft® microfibers are applied for insulation, while a breathable Aquatex™ membrane assures the waterproofness of the system. A heating panel can reach a temperature of 57–63°C, powered by a 7 V or 12 V batteries. Wireless heat control can be used, adjusted by a smart phone application.

Smart clothing for cold protection is also produced by Venture Heat: heated jackets, gloves, and basic line shirts and pants. The incorporated heating panels are made of micro-alloy fibers, powered by rechargeable Li-ion battery. Highly breathable layers from inside and breathable waterproof layers from outside the heating panels are used to assure the body's thermophysiological comfort. The heated gloves of Venture Heat can protect the hands from the cold up to 5 h, depending on the temperature, the applied heating power, and the body's activity. A heated scarf, which can be added to the clothing ensemble, may ensure cold protection from 2 to 6 h. The scarf is made of 100% polyester fibers, and a carbon fiber heating source is placed at the center of the scarf (in the neck zone).

The heated jacket of Flexwarm uses a technology that can add to the efforts of the body's thermoregulatory system to ensure the thermophysiological comfort. Two types of sensors are placed in the jacket: sensors that control the temperature of the body and sensors that detect the temperature of the surrounding air. The heating panels can be separately controlled. Their thickness is equal to the thickness of cotton fabrics, 0.5 mm. The wearer can control the temperature of the panels, which can be heated up to 65°C. Due to the flexible Flexwarm® heating layer, the cold protection garments can be rolled or twisted without a risk of damage.

6. Management of the occupational activities in the cold

The International Standard ISO 15743 [21] deals with the risk assessment and management in a cold environment. It gives both theory and practical tools (checklists, guidelines, examples) to manage the occupational activities in the cold, minimizing the risk of appearance of cold-related injuries. The International Standard ISO 12894 [22] gives details and recommendations on the medical supervision of people, exposed to extreme cold.

The management of the occupational activities in a cold environment helps people to perform their work. Preliminary and regular screening, made by professionals in occupational health, helps to avoid the risk of cold-related injuries due to health limitations of the workers or accumulation of cold strain. Information and training of the workers for activities in a cold environment are inseparable parts of their occupational training. All employees of the company/organization, carrying out activities in the cold, should be trained to identify, access, and manage the risks at work, related to the cold exposure [21].

There are three groups of factors, which are important for the management of the occupational activities in the cold:

- Environmental factors: air temperature, air humidity and wind speed
- Individual factors: activity and cold protective clothing
- Organizational factors: work-rest schedule and adequate shelter

Actually, the second and the third groups of factors have to counterbalance the environmental factors, so as to minimize the cold-related health problems of the workers.

One of the approaches to estimate the severity of the environmental factors is to assess the windchill temperature. It is very appropriate for application in outdoor environment and indoor freezers with fast-moving cold air [23]. The wind speed increases the effect of the low ambient temperature, and the cold, felt by humans, is stronger than supposed for the given air temperature.

The International Standard ISO 11079 [24] gives a tool for designing and management of the activities in cold environments. It defines required clothing insulation (IREQ) index, which allows to predict the necessary clothing insulation to protect the human body in a given cold environment, assessing all environmental factors together with the activity. The IREQ index determined the cold stress at two levels: neutral, which corresponds to the thermophysiological comfort of the human body, and minimal, related to the situation when cold strain already appears, the body is constantly cooling, but the thermoregulatory system of the body still can react and maintain the core body temperature within the desired limits.

Activity is very important in a cold environment, as it produces the necessary heat for warming the core body. However, the contact with cold surfaces increases the heat losses from the body: these can be either handling materials or tools or work in a sitting or kneeling posture on cold surfaces. The exposure to wet environment or precipitation increases the heat losses through the clothing layers as they lose their insulation abilities faster when wet. The inside clothing items and especially the underwear to stay dry during the cold exposure is therefore important: an overheating of the body due to activity in the cold may be likewise dangerous.

The duration of the cold exposure is of crucial importance for the seamless occupational activities in the cold. The American Conference of Governmental Industrial Hygienists (ACGIH) has adopted a schedule for periods of work and breaks, developed by the Saskatchewan Department of Labor in Canada [25]. However, the "work warm-up schedule" is based on the environmental factors only and does not account with the individual factors (clothing and activity). The International Standard ISO 11079 [24] introduces the Duration of Limited Exposure index, which allows to calculate the time of the cold exposure, based on the environmental and personal factors. The Recovery Time index is used to predict the time needed for sheltering, which is enough for the body to warm up before continuing the cold exposure.

7. Conclusions

A number of occupations involve substantial exposure to cold environment, natural or artificial. Working in the cold is related to many adverse effects for the human body, health, and productivity. The knowledge about the thermoregulatory reactions of the body in a cold environment and the associated risks of thermal discomfort, cold strain, and injuries helps workers and employers to make a stand against the cold. The selection of the proper insulating clothing, maintenance of continuous activity, and regular shelter for heating the body are the keys to the comprehensive management strategy for protection of people during occupational activities in the cold.

Author details

Radostina A. Angelova

Address all correspondence to: radost@tu-sofia.bg

Technical University of Sofia, Sofia, Bulgaria

References

[1] Rintamäki H., Parsons K., Limits for cold work, problems with cold work, Proceedings from an International Symposium Held in Stockholm, Sweden, Grand Hôtel Saltsjöbaden, November 16–20, 1997, pp. 72–75.

[2] British standard BS7915, Ergonomics of the thermal environment. Guide to design and evaluation of working practices for cold indoor environments, 1998.

[3] Hatch K.L., Textile science, West Publishing Co., Minneapolis, MN, 1993, p. 26.

[4] Angelova R.A. Textiles and human thermophysiological comfort in the indoor environment, CRC Press, NY, 2016, pp. 11–19.

[5] Simon E., Pierau F.-K., Taylor D.C.M., Central and peripheral thermal control of effectors in homeothermic temperature regulation, Physiological Reviews, 66, 1986, pp. 235–309.

[6] Huizenga C., Hui Z., Arens E., A model of human physiology and comfort for assessing complex thermal environments, Building and Environment, 36(6), 2001, pp. 691–699.

[7] Havenith G., The interaction of clothing and thermoregulation, Exogenous Dermatology, 1(5), 2002, pp. 221–230.

[8] Parsons, K. In: Parsons K.C., (Ed.), Human Thermal Environments, 2nd ed., Taylor & Francis, 2003, Abingdon-on-Thames, UK.

[9] Wissler E.H., Steady-state temperature distribution in man, Journal of Applied Physiology, 16(4), 1961, pp. 734–740.

[10] Fiala D., Psikuta A., Jendritzky G., Paulke S., Nelson D.A., Lichtenbelt W., Frijns A., Physiological modeling for technical, clinical and research applications, Frontiers in Bioscience, 2, 2010, pp. 939–968.

[11] Pozos R.S., Danzl D. Human physiological responses to cold stress and hypothermia, Medical Aspects of Harsh Environments, 1, 2001, pp. 351–382.

[12] Griefahn B., Mehnert P., Brode P., Forsthoff A., Working in moderate cold: a possible risk to health, Journal of Occupational Health, 39, 1997, pp. 36–44.

[13] Hassi J., Rytkonen M., Kotaniemi J., Rintamaki H., Impacts of cold climate on human heat balance, performance and health in circumpolar areas, International Journal of Circumpolar Health, 64(5), 2005, pp. 459–467.

[14] Hassi J., Mäkinen T.M., Abeysekera J., Holmér I., Huurre M., Påsche A., Raatikka V.P., Risk assessment and management of cold related hazards in arctic workplaces: network of scientific institutes improving practical working activities. Institute of Occupational Health, Cold Work Action Program, 2001, Oulu Regional Institute of Occupational Health (ORIOH), Finland.

[15] Chen F., Li T., Huang H., Holmer I., A field study of cold effects among cold store workers in China. Arctic Medical Research 50(Suppl 6), 1991, pp. 99–103.

[16] Roberts D.E., Hamlet M.P., Prevention of cold injuries, Medical Aspects of Harsh Environments, Borden Institute, Washington, 2001, pp. 411–427.

[17] Pozos, R.S., Danzl, D., Human physiological responses to cold stress and hypothermia, Medical Aspects of Harsh Environments, 1, 2001, pp. 351–382.

[18] Pozos R.S., Iaizzo P.A., Danzl D.F., Mills W.T., Limits of tolerance to hypothermia. In: Fregly M.J., Blatteis C.M., (Eds.), Handbook of Physiology. Vol 1, Oxford University Press, New York, NY, 1996.

[19] Angelova R.A., Maintaining the workers comfort and safety in extreme temperatures industrial environment, Proc. of EuroAcademy on Ventilation and Indoor Climate, Course 3 "Industrial Ventilation", Pamporovo, Bulgaria, 18–25 October 2007, pp. 197–205, ISBN 978-954-91681-7-4

[20] Angelova R.A., Phase change materials in intelligent textiles for individually maintained thermal environment, Proc. of EuroAcademy on Ventilation and Indoor Climate, Course 2 "Individually Controlled Environment", Pamporovo, Bulgaria, 8–13 May 2007, pp. 73–78, ISBN: 978-954-91681-5-0.

[21] ISO 15743:2008 Ergonomics of the thermal environment—Cold workplaces—Risk assessment and management.

[22] ISO 12894:2001 Ergonomics of the thermal environment—Medical supervision of individuals exposed to extreme hot or cold environments.

[23] Angelova R.A., The effect of clothing insulation on the thermophysiological comfort of workers in artificial cold environment, Revista Industria Textila, vol. 67, No. 5, 2016.

[24] ISO 11079:2007 Ergonomics of the thermal environment—Determination and interpretation of cold stress when using required clothing insulation (IREQ) and local cooling effects.

[25] Alberta W.S., Best practice—working safely in the heat and cold, Government of Alberta, Alberta, Canada 2009.

Chapter 12

Step-By-Step Procedure and Tools to Reduce Work-Related Stress

Azra Huršidić Radulović

Additional information is available at the end of the chapter

http://dx.doi.org/10.5772/66181

Abstract

Based on available guidelines, protocol, etc., of the EU countries as well as evaluated results of the taken measures, step-by-step procedures are proposed as well as the tools of occupational medicine specialist for the identification of stressors and measures that should be taken to reduce stress at work. European Pact for Mental Health and Well-Being has focused its commitment on workplaces and a necessity to promote work settings, to create the atmosphere for mental health promotion emphasizing reconciliation between work and family life, to introduce a program for stress prevention at work and health promotion in the workplace and to support employment, rehabilitation and return to work of the workers with mental health issues and disorders. Any risk assessment at work, especially of psychosocial risks, requires support of employers and active participation of workers. For that reason, "fight against stress" should become a company's policy. A number of procedures that follow are presented by diagrams that lead through methods of analysis and identification, from selection of measures for stressor reduction to the final evaluation. It is a cycle that ends by defining the term for a new analysis and assessment of psychosocial risks followed by the beginning of another cycle.

Keywords: step-by-step procedure, work-related stress, mental health, risk assessment, workplace prevention, occupational medicine specialist

1. Concepts

According to International Labour Organization [1] and European Commission's Guidance on work-related stress [2], **psychosocial hazards** are interaction among job content, work organization and management, other environmental and organizational conditions, and the employees competencies and needs. **Psychosocial risk** likelihood that psychosocial factors have a hazardous influence on employees' health through their perceptions and experience and the severity of ill health that can be caused by exposure to them. **Work-related stress** pattern of emotional, cognitive, behavioural and physiological reactions to adverse and noxious aspects of work content, work organization, work environment and poor communication. **Burnout** is a state of physical, emotional and mental exhaustion that results from long-term involvement in work situation that is emotionally demanding [3].

Stress is not a disease, but it is the first sign of a problem; if the body experiences a continuous strain, stress can cause acute and chronic changes which can provoke long-term damage to systems and organs, particularly if the body cannot rest and recover. People exposed to stressors have physiological responses. Levels of adrenalin and cortisol hormones consistently rise in response to stress. Those stress hormones effect on blood pressure and cholesterol levels [4]. If chronically repeated, elevation adrenaline and cortisol are likely to have long-term consequences for health.

In the longer period, stress can contribute to hypertension and, as a consequence, to the development of heart and cardiovascular diseases, musculoskeletal disorders, metabolic syndrome and diabetes, as well as peptic ulcers, inflammatory bowel diseases. It can also alter immune functions, which may in turn facilitate the development of cancer. Other than these physiological and physical effects, stress has psychological and social effects. Stress impacts on both affective and cognitive outcomes such as memory loss, attention, decision-making, distress, anxiety, depression, burnout. Exposure to psychosocial risks has been linked to an wide array of unhealthy behaviours like excessive smoking and drinking, physical inactivity, irregular sleep and diet. These disorders are responsible for the great majority of diseases, disability and medical care use, death in most countries.

The society suffers from a loss of work capacity in individual due to direct financial health care costs and decreased quality of life. Work-related stress is determined by psychosocial hazards found in: work organization, work design, working conditions and labour relations.

2. Psychosocial risks factors

Workplaces are constantly evolving following changes in economic and social conditions in society. These new situations pose new challenges. The European Agency for Safety and Health at Work (EU-OSHA)'s second European Survey of Enterprises on New and Emerging Risks (ESENER-2) 2014 provides interesting information on some of the changers. In context of societal change, ESENER-e findings reflect the continues growth of the service sector [5]. The most frequently identified risk factors are interaction with difficult customers, students or patients (58% of establishments in the EU-28), followed by strenuous or painful work posture (56%) and repetitive hand or arm movements (52%). Among 16 most often causes of risk at work in the survey, psychosocial risk was in first place but also positioned at no. 7, as a form of long and irregular working hours and at no. 11 as a form of poor communication,

http://dx.doi.org/10.5772/66181

unsafe work, lack of possibility to influence on work and discrimination against gender, age or nation.

European Social Partners brought Framework Agreement on Work-related Stress in Brussels 2004. "The agreement" provides an action-oriented framework [6]. This includes non-exhaustive examples of **indicators, risk factors and measures** that should guide action.

-Suggested indicators for the **identification** of stress-related problems are high absenteeism, high staff turnover, frequent interpersonal conflicts or complaints by workers.

-The agreement provides an **optional list of risk factors or stressors** to be analysed:

- **Work organisation and processes** (working time arrangements, degree of autonomy, match between workers' skills and job requirements, workload, etc.)
- **Working conditions and environment** (exposure to abusive behaviour, noise, heat, dangerous substances, etc.)
- **Communication** (uncertainty about what is expected at work, employment prospects, or forthcoming change such as restructuring or new technologies/processes, etc.) and
- **Subjective factors** (emotional and social pressures, feeling unable to cope, perceived lack of support, etc.).

-Regarding concrete **measures** to take, the agreement leaves some flexibility: "preventing, eliminating or reducing problems of work-related stress can include various measures. These measures can be collective, individual or both. They can be introduced in the form of specific measures targeted at intensified stress factors or as part of an integrated stress policy encompassing both preventive and responsive measures".

Nowadays, most EU countries adopted a list psychosocial hazard of six key areas of work design: **demands, control, support, relationships, role and change** [7].

Guidance on the management of psychosocial risks in the workplace PAS 1010:2011 was published in 2011 [8]. It elaborates in detail some of key issues:

- Job content–Lack of variety or short work cycles, fragmented or meaningless work, underuse of skills, high uncertainty, continuous exposure to people through work
- Workload and work pace–Work overload or underload, machine pacing, high levels of time pressure, continually subject to deadlines
- Work schedule–Shift working, night shifts, inflexible work schedules, unpredictable hours, long or unsociable hours
- Control–*Low participation* in decision-making, lack of control over workload, pacing, shift working
- Environment and equipment–Inadequate equipment availability, suitability or maintenance, poor environmental conditions such as lack of space, poor lighting, excessive noise

- Organizational culture–Poor communication, low levels of support for problem-solving and function personal development, lack of definition of, or agreement on, organizational objectives
- Interpersonal relationships–Social or physical isolation, poor relationships with superiors, interpersonal at work conflict, lack of social support, harassment, bullying, third-party violence
- Role in organization–Role ambiguity, role conflict, responsibility for people
- Career development–Career stagnation and uncertainty, under-promotion or over-promotion, poor pay, job insecurity, low social value to work
- Home-work interface–Conflicting demands of work and home, low support at home, problems relating to both partners being in the labour force (dual career).

A large number of data were obtained by employees' stress self-assessment, according to which 60% of the participants regard responsibility due to their line of work as stressful; 52% participants consider stressful situation interruptions at work; for 46% of employees, stressful situations are short notice deadlines; for 45% of employees, repetitive movements are stress inducers; and for 40% of employees, new tasks are stressful, especially for older working population [9]. A number of questionnaires are offered for self-assessment of the increased stress results, that is, burnout syndrome. In the Netherlands, 22% of the cases of burnout syndrome were caused by a job organization (too few or too many tasks, unclear duties, etc.), while the interpersonal relations at work hold the second place (21%) [10].

In the Netherlands, specifically among most stressful jobs is work behind a counter followed by the job of a teacher [10].

3. Mental disorders, work and cost

With regard to mental disorders in working population, three big groups of patients can be recognized. The first group of employees is the ones who have work-related mental health problems, while the second one has similar problems that are not work-related. Workplace is of great importance to both groups because of their rehabilitation process. The third group includes the unemployed suffering from mental disorders for a longer period of time, but it is also important to integrate them in the working process [11].

"European Pact for Mental Health and Well-being of 2008" gives workplace a central role underlying the necessity for improvement in workplace setting by creating a good work environment to contribute to mental health well-being by including reconciliation of work and family life, introducing stress prevention program at work, promoting health at workplace and support to employment, rehabilitation and return to work of employees with mental problems and mental ill health [12]. Luxembourg Declaration points out the importance of workplace health promotion since a considerable number of population spend significant amount of time at work. To an individual being employed is an indicator of a social status, life satisfaction and sense of purpose which contributes to a person's health. There are algorithms for workplace health protection and promotion that can be used for employee's [13]. Occupational medicine specialist as a physician who does preventive health check-ups [14] can discover acute problems and address them, demand intervention at an individual or group level and implementation of efficient measures of monitoring and checking up [15]. In the Benelux Union and France, occupational medicine specialist has legal duty to annually analyse psychosocial risks [16].

To which extent a good communication, promotion at work and praise, positive attitude to work and colleagues as well as the feeling of giving (in one word well-being) contribution to the organization business are important is illustrated by the fact that employees who do not have positive attitude towards work and colleagues spend 3.5 days longer on sick leave every year. It is particularly important to emphasize that only 56% employees consider their income to be main motive for work, and it is stimulation only when 26% above the average [17]. The results of the survey showed that only 20% people successfully solved the problem if deconcentrated by offensive remarks or disturbed in some other manner, while 80% solved the same task if positively stimulated and praised. Humour at work reduces stress and improves the feel-good factor closer connecting the working community. Bad relationship in the workplace caused depression in 30% individuals [10].

"*GreenPaper*" of European commission–improving the mental health of the population–towards a strategy on mental health for the European Union of 2005 claims that mental health costs the EU an estimated 3–4% of GDP, mainly through lost work productivity. Furthermore, up to 28% of employees in Europe report stress at work [18]. The fact that mental health and the mental health-related problems affect 38.2% of European population every year illustrates how serious public health problem it is [19].

The Fourth European Working Conditions survey (Eurofound, 2007) found that work-related stress alone affects more than 40 million individuals across the European Union, costing estimated Ĩ 20 billion a year in lost time and health bills; it is the most commonly reported causes of occupational illness by workers [20].

Studies show profitability in investing funds in reduction stressrisk at work. For each invested Euro in reduction, from 0.81 to 13.62 Euros are returned to the investor [21]. The research results from 2013 submitted to the US Congress emphasize that the return of the invested money in stress reduction is even higher in the second and the third years after the investment [22].

4. Step-by-step procedure

Chart gives a detailed review of the step by step representing a number of procedures conducted in a company in order to assess psychosocial risks and select measures for risk reduction in a workplace. Step by step was created on the basis of available directives, algorithms of the procedure, experience in antistress procedure implementation in some developed countries and evaluation of the results of the measures taken, adjusted to tradition, resources and legal provisions, published in Croatian [23] and presented on Final Conference of the EU Joint Action on Mental Health and Well-being (JA MH-WB) [24]. Based on available guidelines, protocol, etc., of the EU countries as well as evaluated results of the taken measures step-by-step procedures are proposed as well as the tools of occupational medicine specialist for identification of stressors and measures that should be taken to reduce stress at work.

The most important is the implementation of risk-at-work assessment, selection of measures and their control, primarily by employer along with employees and the assistance of an expert/employee in charge of safety at work and occupational medicine doctors/specialists [7, 14, 15, 25, 26].

The developed world and Europe apply several strategies and models that are focused on organization's analysis and its existing activities, mainly those with regard to prevention and safety at work, health conditions of employees, selection of the most suitable tools for identifying psychosocial risks and stressors. One of the best known for all types of workplace risk assessment is SOBANE strategy which includes four levels: screening, observation, analysis and expertise level

(*S–.screening, OB–.observation, AN–.analysis and E–.expertise*) [27]. With regard to the stress level according to Karasek's model (demand/control), organizations are divided into four types: active (high job demands and high control), passive (low job demands/low control), high strain (high demands/low control) and organizations with little strain (low demands/high control) [28].

Each workplace risk assessment especially psychosocial risk assessment, requires employer's support and employees' active participation. Therefore, "fight against stress" should become **organization's policy** (the first step–**Figure 1**). Along with the awareness of the necessity to fight against stress and employer's support, it is necessary to motivate the employers and employees alone to participate in the workplace psychosocial risk assessment, as the primary goal is satisfaction of employees at their work. Healthy workplace environment and satisfaction at work have as a result a higher productivity and a better quality work. According to the latest indicators for organization's management in fighting stress, the most important motive is complying with legal duties, meeting the employees' and their representatives' expectations avoiding the inspection fines but also maintaining the organization's reputation and increase in its productivity [5].

The second step is establishing of a **committee for stress reduction in the workplace, that is**, a team to identify and analyse indicators, risks stressors, identify groups or individuals exposed to high risks, the so-called focus groups and select and propose measures for psychosocial risk reduction at work and at the same time, stressreduction. The committee carries out an action plan and makes decisions about measures to take, who takes measures, how the action is financed, who is responsible and the time needed for the action to be done. So, the owner and/or employer, that is, the authorized person is the key member of the committee for stress reduction in the workplace.

http://dx.doi.org/10.5772/66181

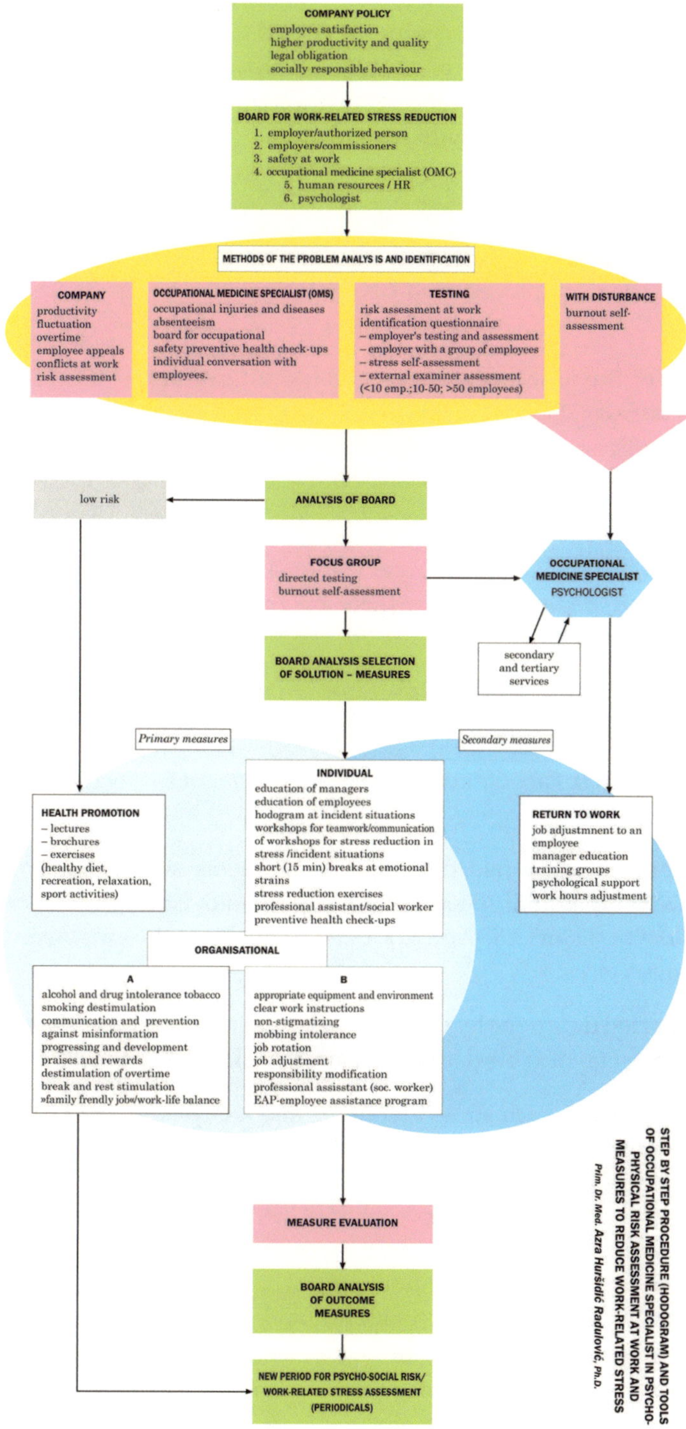

Figure 1. Step-by-step procedure and tools to reduce work-related stress.

Occupational medicine specialist is obliged to assess working environment and to work on risk prevention which is in accordance with the signed Convention 161 of *International Labour Office* (ILO) [29]. Employer is due to provide occupational medicine service and health monitoring corresponding with dangers, risks, health damages and strain at workplace with the aim of protecting employees' health.

Furthermore, in some countries have legally established the Committee for Safety and Health Protection and its members including: employers or employer's authorized person, representatives of employees, an expert for safety at work and an occupational medicine specialist [30]. Personnel administration employees and psychologists may be also included in the work of the committee.

Different **methods of the problem analysis and identification** (the third step) are applied in business analysis and work so-called stress indicators, not only of the organization but also of the occupational medicine specialist. Risk assessment of an organization is a legal obligation for employers and has a particular importance in getting the insight into safety and protection at work.

Along with the data obtained from an organization and obligatory riskassessment, data and occupational medicine specialist findings, there are a number of tests to enable a better insight into psychosocial risks in the workplace.

Employer alone can identify work stressors of his employees by means of work stressor questionnaire but there is a possibility for external assessor to do that.

A number of questionnaires are offered for self-assessment of the increased stress results, that is, burnout syndrome.

Employees who have already had disorders or problems should be sent to occupational medicine specialist who would, together with a psychologist as a member of occupational medicine team, take necessary measures, for example, direct the employee to the secondary and tertiary health services.

Analysis of all collected data done by the Committee members (the fourth step) can have many outcomes and the most favourable one is a **low stress-risk**, when no measures should be taken.

A group or individuals in which stress identifiers and analysis showed a high stress risk, the so-called **focus group** will probably require additional analysis and testing as well as interviews with high stress risk employees. Some individuals will need occupational medicine specialist service, in some cases even psychologist, which is secondary, that is, reactive action against stress in case of an individual under stress or exposed to stressrisk. Primary or proactive action against stress at work cannot be and need not be removed since in a lower degree it can positively affect creativity, quality, concentration and productivity. Usual steps in removing the danger, harmfulness and strain at work are undoubtedly their elimination, reduction, isolation, control, information and consultation.

After analysis, the committee chooses the measures (the fifth step) for stressogenic factors reduction according to the so-called SMART criteria (*S–specific*; *M–measurable*; *A–attainable*; *R–*

realistic; T–time-bound) [31]. EU studies point out that the measures taken to reduce psycho-physiologic strain in the workplace and health promotion that result in changes of lifestyle increase productivity and profitability and reduce fluctuation as such workplaces are regarded as the "chosen workplaces" [21]. Health promotion at work is a *primary measure for psychosocial risk reduction in the workplace*. European network for health promotion at work gives support to professionals and non-professionals in the field, at the national level [32].

A-organizational measures are primary measures for stress reduction at work. Employer and managers should stimulate and be a role model for using a holiday and break at work, discourage overtime and insist on reconciliation of family life and work. Bigger organizations invest their money in building kindergartens and its equipment, centres for physical training, which, along with health promotion, have positive impact on employees' work.

B-organizational measures are primary and secondary measures for stressreduction in the workplace. Strictly speaking "organizational measures" refer to a suitable workplace, work conditions and methods of work of the employees, for example, working hours, work post, lighting, noise and clear work instructions. Bigger organizations in developed European and US states have introduced the employee assistance program (EAP). It enables the employees with the family problems and problems at work to turn to professional assistants for help for free [33].

Individual measures include changing the attitude, methods of work and code of conduct. The contribution of management is seen in education of employees in conducting a conflict, team building and education of team leaders. In Denmark, managers have longer sick leaves and more claims for recognition of professional stress [10]. Education and exercises teach the employees the techniques of stress reduction, relief and relaxation. Efficiency of individual measures for absenteeismreduction is proved [34]. Considerable number of jobs, especially in growing service industry, require emotional strain at work. These jobs require "empathy and sensibility", and emotional stress does not end with finishing work [33, 35]. They are jobs which cause frustration and the feeling of helplessness, jobs in which the employees are faced with extreme situations, so they need some time to move away from their job. Traumatic events like accidents and attacks in the workplace demand special measures starting from step-by-step procedure for incidents. It was found that traumatic experience caused post-traumatic stress syndrome in 87% of the employees [10]. Individual techniques for stressreduction and frequent 15 min breaks at work to do some physical exercises turned to be very efficient [33]. Individual measures at work help in reduction of home-related stress, such as divorce, moving a house and death of a close relative. Not only employers with mental illnesses or disorders should be included in the working process regardless of the cause being at work or outside work, but also the unemployed with chronical mental disorders.

Return to work of the employees with mental illness or disorders requires a number of adjustments, from working hours to education of managers and from trainings of the group of the employees to which such employee would join, to psychological support.

Evaluation of measures (the sixth step) will show if the committee has selected and worked out suitable measures and if their implementation has led to the desired goal, that is, to the reduction in psychosocial risks at work and reduction in stress in the workplace or work-related stress.

Committee analysis of the results of the measures taken the seventh step should answer the question of profitability of the measures.

All those results will define **a new term to start evaluation of psychosocial risks/stress at work** (the eighth step).

Tradition and experience in safety at work of occupational medicine specialist are exceptionally powerful element in fighting stress at work, which is recognized in EU project "*Jointaction* WP6-mental health at work (personal data)". Also, it should be mentioned that mental health is obligatory part of education of occupational medicine specialist. Considering all obtained results and experience from other countries, we can say that by implementing all mentioned methods, measures and procedures for stress reduction at work and the acquired experience as well as education and evaluation of all results and within the existing legal framework, occupational medicine specialist can make a big step forward towards the stress reduction at work and achieve its primary goal.

Author details

Azra Huršidić Radulović

Address all correspondence to: ordinacija-medicine-rada@zg.t-com.hr

Occupational Medicine Practice, Zagreb, Croatia

References

[1] International Labour Organization (1984). Psychosocial factors at work: recognition and control. Occupational Safety and Health Series No: 56. Available from: http://www.who.int/occupational_health/publications/ILO_WHO_1984_report_of_the_joint_committee.pdf

[2] European Commission (2002). Guidance on work-related stress–.spice of life or kiss of death? Luxembourg: Office for Official Publications on the European Communities.

[3] Schaufeli W.B., Greenglass E.R. (2001). Introduction to special issue on burnout and health. Psychology and Health. 16; 501–510.

[4] WHO (2010). Health impact of psychosocial hazards at work: an overview. Available from: http://www.who.int/occupational_health/publications/hazardpsychosocial/en/

[5] Second European Survey of Enterprises on New and Emerging Risks (ESEN-ER-2)–.overview report: managing safety and health at work 2014. Available from:

https://osha.europa.eu/en/tools-and-publications/publications/second-european-survey-enterprises-new-and-emerging-risks-esener/view

[6] European Social Partners (2004). Framework agreement on work-related stress. Brussels: European Social Partners–ETUC, UNICE, BUSINESSEUROPE, UEAPME and CEEP. Available from: http://www.europarl.europa.eu/registre/docs_autres_institutions/commission_europeenne/sec/2011/0241/COM_SEC(2011)0241_EN.pdf

[7] Health and Safety Executive (HSE) (2007). The suitability of HSE` risk assessment process and management standards for use in SMEs (RR537). Available from: http://www.hse.gov.uk/research/rrpdf/rr537.pdf

[8] The British Standards Institution (BSI) (2011). Guidance on the management of psychosocial risks in the workplace PAS 1010:2011. Available from: http://shop.bsigroup.com/en/ProductDetail/?pid=000000000030213276

[9] Lohmann-Haislah A. [Stressreport Deutschland 2012, in German]. Available from: http://www.baua.de/de/Publikationen/Fachbeitraege/Gd68.

[10] What recognition of work-related mental disorders? A study on 10 European countries study Report Eurogip 2013. Available from: http://www.europeanforum.org/documents/24/eurogip_81en_recognition_wr_mental_ disorders_europe.pdf

[11] Fine A, Griffiths J, editors. Joint action on mental health and well-being. Mental health at the workplace. The SWOT analysis report–context and key findings 2015. Available from: http://www.eaad.net/fileadmin/user_upload/06_JA_SWOT_Analysis_Report_Part_1.pdf

[12] European Pact for Mental Health and Well-being "Together for Mental Health and Well-Being" EU High Level Conference; 12–13. June 2008, Brussels.

[13] The Luxembourg Declaration on Workplace Health Promotion in the EuropeanUnion (2007). Available from: http://www.enwhp.org/fileadmin/rs-dokumente/dateien/Luxembourg_Declaration.pdf

[14] HSPS 15 Stress and Resilience (2013). Health, Safety and Wellbeing. Version 5. Available from: http://www.plymouth.gov.uk/hsps15_stress_and_resilience.pdf

[15] [Direction générale Humanisation du travail. Methodes et instruments pour une analyse ergonomique et psychosociale, in Franch]. (2005). Bruxelles: Directionegénérale Humanisation dutravail.

[16] Langevin V, Francois M, Boini S, Riou A. (2011). [Évolutions et Relations en Santé au travail (EVREST), fran.]. Doc Méd Trav 127: 463–5.

[17] [Studie zur Job-Motivation zeigt,was Fach-und Führungskräfte in Deutschland zu mehr Leistung antreibt]. (2012). Düsseldorf. Available from: http://www.stepstone.de/Ueber-StepStone/presse/motivation-ist-nicht-kaeuflich.cfm

[18] Health and Consumer Protection Directorate-General (2005). GreenPaper, Improving the mental health of the population: towards astrategy on mental health for the

European Union, Brussels. Available from: http://www.ec.europa.eu/health/archive/ph_determinants/life_style/mental/green_paper/mental_gp_en.pdf

[19] Wittchen H.U., Jacobi F., Rehm J., Gustavsson A., Svensson M., Jönsson B., Olesen J., Allgulander C., Alonso J., Faravelli C., Fratiglioni L., Jennum P., Lieb R., Maercker A., van Os J., Preisig M., Salvador-Carulla L., Simon R., Steinhausen H.C. (2011). The size and burden of mental disorders and other disorders of the brain in Europe 2010. Eur Neuropsychopharmacol 21: 655–679. doi:10.1016/j.euroneuro.2011.07.018

[20] EuroFound (2007). The Fourth European Working Conditions. Dublin: Office for Official Publications on the European Communities. Available from: http://www.eurofound.europa.eu/publications/report/2007/working-conditions/fourth-european-working-conditions-survey

[21] Economic analysis of workplace mental health promotion and mental disorders prevention programmes and of their potential contribution to EU health, social and economic policy objectives (2013). Available from: http://ec.europa.eu/health/mental_health/docs/matrix_ economic_analysis_mh_promotion_en.pdf

[22] U.S. Department of Health & Human Services (2013). Report to congress on workplace wellness. Available from: http://aspe.hhs.gov/hsp/13/WorkplaceWellness/rpt_wellness.cfm

[23] Huršidić Radulović A. (2015). [Hodogram specijalista medicine rada za procjenu psihosocijalnih rizika i mjere za smanjenje stresa na radu]. Arh Hig RadaToksikol. 66: 85–90

[24] Final Conference of the EU Joint Action on Mental Health and Wellbeing (JA MH-WB), (21–22 January 2016), Brussels, Belgium. Promoting mental health at workplaces. Available from: http://www.jamhwbfinalconference.admeus.net/uploads/jamhwbfinalconference/editor/Step%20by%20step%20%20-%20Azra.pdf) and http://www.jamhwbfinalconference.admeus.net/?page=48

[25] [Valutatione e gestione del rischio da stres lavoro correlato. Manuale ad uso delle aziende in attuazione del D.Lgs.81/08e s.m.i. INAIL 2011, in italien]. Available from: http://www.inail.it/internet_web/wcm/idc/groups/internet/documents/document/ucm_portstg_093254.pdf

[26] Federal Ministry of Labour and Social Affairs (2013) Joint declaration on mental health in the workplace. Bonn: Federal Ministry of Labour and Social Affairs.

[27] Malchaire J.B. (2004) The SOBANE risk management strategy and the Déparis method for the participatory screening of the risks. Int Arch Occup Environ Health 77: 443–450. PMID: 15205963

[28] Karasek's Model of Job Strain (R.A. Karasek,1979). Available from: http://www.med.uottawa.ca/courses/epi6181/Course_Outline/Karasek_fn.pdf

[29] C161–.Occupational Health Services Convention (1985). (No. 161). Available from: http://www.ilo.org/dyn/normlex/en/f?p=NORMLEXPUB:12100:0::NO::P12100_INSTRUMENT_ID:312306

[30] [Zakono zaštiti na radu. Narodne novine 71/2014, in Croatian]. Available from: http://narodne-novine.nn.hr/clanci/sluzbeni/2014_06_71_1334.html

[31] Doran G.T. (1981). There's a S.M.A.R.T. way to write management's goals and objective. Manag Rev (AMA FORUM). 70: 35–36.

[32] The European Network for Workplace Health Promotion (ENWHP) 2001. Available from: http://www.enwhp.org

[33] PsychischeGesundheitinderArbeitswelt–psyGA 2015. Available from: http://psyga.info/ueber-psyga/das-projekt/.

[34] Sockolt I, Kramer I, Badecker W. Effectiveness and economic benefits of workplace health promotion and prevention. iga-Report 13e 2009. Available from: http://www.iga-info.de

[35] Travis D.J., Lizano E.L., Mor Barak M.E. (2016). 'I'm So Stressed!': a longitudinal model of stress, burnout and engagement among social workers in child welfare settings. Br J Soc Work. 46(4): 1076–1095. Epub 2015 Mar 4.